Biochar Systems for Smallholders
in Developing Countries

A WORLD BANK STUDY

Biochar Systems for Smallholders in Developing Countries

Leveraging Current Knowledge and Exploring Future Potential for Climate-Smart Agriculture

Sebastian M. Scholz, Thomas Sembres, Kelli Roberts, Thea Whitman, Kelpie Wilson, and Johannes Lehmann

THE WORLD BANK
Washington, D.C.

Contents

Figures

Tables

Acknowledgments

This report was written by the following team of authors: Sebastian M. Scholz (senior environmental economist, World Bank), Thomas Sembres (programme officer, UNEP UN-REDD Programme), Kelli Roberts (postdoctoral associate, Cornell University), Thea Whitman (doctoral researcher, Cornell University), Kelpie Wilson (project development director, International Biochar Initiative), and Johannes Lehmann (associate professor, Cornell University). The World Bank task team, led by Sebastian M. Scholz, also included Gerhard Dieterle (forests adviser) and Ellysar Baroudy (senior carbon finance specialist). In addition, very constructive peer-reviewed feedback was received from members of the World Bank's Latin America and Caribbean unit as well as from Sameer Akbar from the Climate Change Group. The authors thank Karin Kemper, director of the Climate Change Group, Climate Change Policy & Finance, who formerly managed the Latin America and Caribbean Environment team. The authors also thank Ede Jorge Ijjasz-Vasquez, sector director of the Latin America and Caribbean Sustainable Development Department at the World Bank. Both were instrumental in providing ongoing advisory support throughout the making of this report. Robert J. Griffin, with the World Bank's BioCarbon Fund, was instrumental in managing the report through publication and reviewing all research from contributing authors.

The report benefited from the review comments provided by Ian Noble (chief scientist, Global Adaptation Institute) and Erick C.M. Fernandes (adviser, World Bank). Comments were also received from Rosina Bierbaum (professor, University of Michigan) and Christian Witt (senior program officer, Soil Health, Bill & Melinda Gates Foundation). The latter two reviewers were also members of a group of external experts who provided valuable insights and constructive advice during the preparation of this report. Other members of the external guidance group included Pedro Sanchez (director, Columbia University), Sasha Lyutse (policy analyst, NRDC Center for Market Innovation), and Patricia Elias (forest science and policy consultant, Union of Concerned Scientists).

The report was strengthened through discussions during a workshop held in Washington, DC, on May 11, 2011. Comments and advice from the workshop participants helped to improve the report.

A number of biochar practitioners and researchers from around the world provided project-related information in response to a global biochar systems survey, undertaken as part of this study. Their participation is gratefully acknowledged, as is access to this network of practitioners that the International Biochar Initiative provided.

Core funding for this undertaking was provided by the Program on Forests and the World Bank's BioCarbon Fund. Additional funding was made available by the Carbon War Room.

About the Authors

Johannes Lehmann is an associate professor of soil biogeochemistry and soil fertility management at Cornell University. He received his graduate degree in soil science at the University of Bayreuth, Germany. During the past 10 years he has focused on nanoscale investigations of soil organic matter, the biogeochemistry of black carbon, and sequestration in soil. Doctor Lehmann is a member of the steering group of the U.S. National Soil Carbon Network. He has testified in the United States Congress and briefed the President's Council of Advisors on Science and Technology. He was part of Workgroup 2 on Monitoring and Assessment of Sustainable Land Management of UNCCD, and is on the editorial boards of several international journals.

Kelli Roberts is a postdoctoral associate at Cornell University in the Department of Crop and Soil Sciences. She completed her PhD at the University of Washington in Seattle and her undergraduate in materials science and engineering at the Massachusetts Institute of Technology. Her research has focused on using life cycle assessment and systems thinking to analyze biochar systems. She also has a strong interest in sustainable agricultural systems and has established a small organic vegetable farm where she lives in Rhode Island.

Sebastian M. Scholz is a senior environmental economist. He works mainly in the area of climate finance/carbon finance and climate change mitigation for the Sustainable Development Network of the World Bank. His current regional focus is on Latin America and the Caribbean; he also provides cross-support to other regions, particularly on climate mitigation activities. He has a PhD in agricultural sciences and agricultural economics and an MSc in forestry and natural resource management from the Justus-Liebig-University in Giessen, Germany, and the Technical University of Munich, Germany, respectively.

Thomas Sembres is an economist working at the UN-REDD Programme in the United Nations Environment Programme in Nairobi. He works at the policy level on strategies and investments to reduce deforestation and forest degradation in the Congo basin. Prior to joining the UN-REDD Programme, Thomas worked at the World Bank's Forest Team and Program on Forests in Washington, DC. He also worked as adviser on opportunities to reduce deforestation and forest degradation in Central and East Africa for the Forests Philanthropy Action Network, based in London, United Kingdom. He holds a Master's degree in environmental policy from Cambridge University.

Thea Whitman is a PhD graduate student at Cornell University in the Department of Crop and Soil Sciences. She completed her MS at Cornell in 2011, focusing on biochar and climate change, including laboratory research on biochar carbon stability, system dynamics modeling of the climate change impact of a biochar cookstove system in western Kenya, and an investigation into the policy issues surrounding biochar projects for climate change mitigation. Her current research focuses on the interactions between biochar and nonbiochar soil organic matter, microbes, and plants, in a climate change context. She also has an avid interest in international climate change policy and has attended the past three United Nations climate change negotiations (COP14–COP16) as a member of the Canadian Youth Delegation.

Kelpie Wilson is a project development director at International Biochar Initiative. She is a writer and mechanical engineer with experience in journalism, technology assessment, community organizing, forest policy, and NGO management. She has worked for IBI on communications and project development since 2008. Prior to that, she was the environmental editor and columnist for Truthout.org and a contributor to numerous online and print publications. In the 1990s, she was the executive director of the Siskiyou Regional Education Project, a grassroots environmental group protecting wilderness and old growth forests in the Siskiyou Mountains of Oregon. She has a BS (with honors) in mechanical engineering from California State University, Chico. From 1986 to 1989 she worked for several firms doing research and development on Stirling cycle engines, including an internship at the Jet Propulsion Laboratory in Pasadena, California. She has also worked as a technical writer for a solar power company.

Abbreviations

AEC	anion exchange capacity
BC	black carbon
BEGGE	biochar energy, greenhouse gases, and economic
BioCF	BioCarbon Fund
C	carbon
CDCF	Community Development Carbon Fund
CDM	Clean Development Mechanism
CEC	cation exchange capacity
CH_4	methane
C:N ratio	carbon to nitrogen ratio
CO_2e	carbon dioxide equivalent
FAO	Food and Agriculture Organization
fNRB	fraction of nonrenewable biomass
GEF	global environment facility
GHG	greenhouse gas
GREET	greenhouse gases, regulated emissions, and energy use in transportation (model)
IBI	International biochar initiative
IPCC	Intergovernmental Panel on Climate Change
K_2O	potassium oxide
LCA	life-cycle assessment
LED	light emitting diode
MRT	mean residence time
NGO	nongovernmental organization
NMHC	nonmethane hydrocarbons
N_2O	nitrous oxide
NPK	nitrogen, phosphorus, and potassium (fertilizer)
NRDC	Natural Resources Defense Council
O:C ratio	Oxygen to carbon ratio

OECD	Organisation for Economic Co-operation and Development
PAH	polycyclic aromatic hydrocarbon
$PM_{2.5}$	particulate matter less than 2.5 micrometers in diameter
P_2O_5	phosphorus pentoxide
REDD	reducing emissions from deforestation and forest degradation
SImpaCCT	Stove Impact on Climate Change Tool
SOC	soil organic compound
SOM	soil organic matter
TLUD	top-lit updraft
UNCCD	United Nations Convention to Combat Desertification
UNEP	United Nations Environment Programme
UN-REDD	United Nations–Reducing Emissions from Deforestation and forest Degradation
VND	Vietnamese dong

All dollar amounts are U.S. dollars unless otherwise indicated.

Executive Summary

Introduction

Three of the biggest challenges of the twenty-first century are the need to nearly double food production by 2050, to adapt and build resilience to a more and more challenging climatic environment, and to *simultaneously* achieve a substantial reduction in atmospheric greenhouse gas concentrations. The surge of interest in climate-smart agriculture, which focuses on solutions to the three challenges, has sparked curiosity in using biochar as a tool to fight climate change while also improving soil fertility. Biochar systems are particularly relevant in developing country contexts and could be leveraged to address global challenges associated with food production and climate change. However the potential effects of biochar application to soils are diverse and its climate impact is contingent on the design of the system into which it is integrated. Thus biochar systems are inherently complex and further research is needed to understand their associated opportunities and risks in developing countries.

There are a number of reasons why biochar systems might be particularly relevant in developing-country contexts. The potential for biochar to improve soil fertility could result in increased crop yields from previously degraded soils for smallholder farmers. Improved cookstoves that produce biochar as well as heat for cooking could reduce indoor air pollution and time spent on fuel gathering. Both of these results could be beneficial to forests. Enhanced food production capacity could potentially decrease the need to clear more forested land for agriculture, and more efficient cookstoves could decrease wood gathering from forests already in decline. It is vital that further research is undertaken to fill the gaps in our knowledge of biochar systems.

This report offers a review of what is known about opportunities and risks of biochar systems including soil and agricultural impacts, climate change impacts, social impacts, and competing uses of biomass. The report benefited from its wide-ranging methodology including a desk review of existing literature; a two-step survey of biochar systems that elicited 154 responses to the initial survey, and 48 responses on the follow-up survey to learn more about the social and cultural barriers to biochar adoption; an expert workshop in Washington, DC to assist in analysis of the collected data; development of a typology of biochar systems; and a life-cycle assessment of selected systems in Kenya, Vietnam, and Senegal.

Background on Biochar

Biochar is the solid product remaining after biomass is heated to temperatures typically between 300°C and 700°C under oxygen-deprived conditions, a process known as "pyrolysis." Biochar is a system-defined term referring to black carbon that is produced intentionally to manage carbon for climate change mitigation purposes combined with a downstream application to soils for agricultural effects. It is produced with the intent to be applied to soil as a means of improving soil productivity, carbon storage, or both. Although the term "biochar" has come into common usage only relatively recently, the practice of amending soils with charcoal for fertility management goes back millennia. Instances can be found in Africa, Asia, and notably in the Amazon basin where the historically managed *terra preta*, or "dark earths," stand out for their capacity to store carbon.

Biomass typically contains about 50 percent carbon, which is relatively quickly decomposed and reemitted to the atmosphere upon decay in soil. The mean residence time of fresh biomass is in the range of months to years, with longer times for woody biomass and colder climates. Biochar retains between 10 percent and 70 percent (on average about 50 percent) of the carbon present in the original biomass and slows down the rate of carbon decomposition by one or two orders of magnitude, that is, in the scale of centuries or millennia.

In principle biochar can be made from any type of biomass. It is important to understand how different production conditions can result in different types of biochars and how these chars interact with different types of soils. Three elements critical to every biochar system are (a) the source of biomass, (b) the means of biochar production, and (c) whether and how it is applied to soil (figure ES.1).

For each element there are a wide range of alternatives. The source, or feedstock, can be almost any type of biomass including agricultural wastes, rice husks, and urban green waste. However focus is placed on use of "true wastes" in order to minimize disruption to local carbon and nutrient recycling. Production systems range in scale from small household cookstoves to large industrial pyrolysis plants. Systems are generally classified as either gasifiers or pyrolysers depending on the technology applied. Several options exist for the use of biochar system outputs. This report focuses on integrated bioenergy-biochar-soil application systems with benefits for soil health and productivity, farming economies, climate, and human well-being.

Figure ES.1 Biochar as a System-Defined Concept

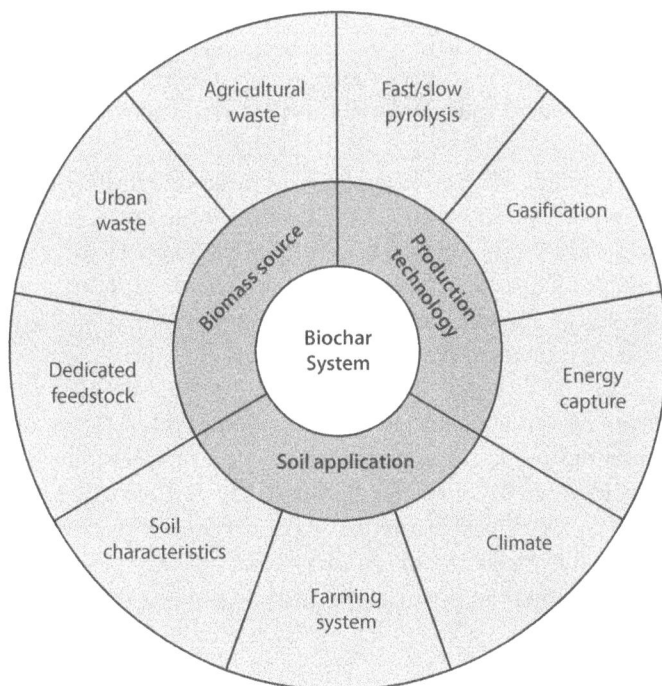

Source: World Bank.

Overall Opportunities and Risks of Biochar Systems

When considering the overall opportunities and risks of biochar systems, four main factors need to be taken into account: the impacts on soil health and agricultural productivity; the impacts on climate change; the social impacts; and competing uses of biomass.

Impacts on Soil Health and Agricultural Productivity

A number of benefits have been reported for biochar application to soils including greater yields, better predictability of yields, reduced germination time, extension of cropping season, and greater resilience to drought. However, the impact of biochar application by location and over time can vary given the range of factors involved including:

- **Soil pH.** Most biochars have a neutral to basic pH, and thus have a liming effect in acidic soils, which increases plant productivity. Application of biochar to basic soils does not necessarily lead to increased productivity.
- **Nutrient availability.** Biochar is not a fertilizer per se but a soil conditioner with the potential to improve soil functions by reducing nutrient loss, increasing bioavailability of nutrients, and decreasing nutrient leaching. Thus, the application of biochar can reduce the reliance on fertilizer and its associated environmental and economic costs.

- **Soil moisture.** Due to its porous structure and large surface area, biochar can retain plant-available water and improve the overall water-holding capacity of soils (with the exception of clay soils).
- **Soil organic matter.** The role of biochar in increasing and sustaining organic matter content in soil is related to the soil type and the timescale over which biochar is applied.
- **Amount of biochar applied.** Studies have shown a positive correlation between increasing levels of biochar addition and plant response but only up to a certain point. Further research is needed to clarify the biophysical optimum for each soil–crop system. The emerging body of literature indicates that there is substantial potential for biochar application to improve crop yields of smallholder, subsistence farmers in developing countries.

Given the complexity of systems and factors and the many unknowns surrounding optional conditions for securing predictable positive effects from biochar application there is great need for further research and pilot projects. Initially it may be useful to focus on soils that have low organic matter content. The abundance of marginal or degraded soils in many parts of the tropics indicates that biochar has great potential to impact agriculture in these regions.

Impacts on Climate Change

The relevance of biochar for climate change mitigation has caught as much attention as its potential for agronomic purposes. The biochar material itself is a carbon concentrate of the original biomass feedstock, and its stability in soil thus has implications for soil carbon sequestration. Factors relevant to the potential climate change impacts of biochar include:

Direct and Indirect Sources of Emission Reductions

Carbon storage and stabilization is probably the most direct and important quality for climate change mitigation efforts based on biochar. The crucial calculation here is the *difference* between the decomposition of the biochar and the decomposition of the original feedstock that would have happened in the absence of pyrolysis. The rate of decomposition of the biochar and consequently its capacity for carbon storage depends on several factors. Two main factors include the ratio of labile carbon (which is readily degradable) to recalcitrant carbon (which is more resistant to degradation), and on the pyrolysis technique used to produce the biochar. Indirect factors contributing to emission reductions include generation of renewable energy through combustion of the syngas by-product of biochar production; waste diversion, thus avoiding methane and nitrous oxide emissions from decomposing wastes; reduced fertilizer manufacturing, if biochar proves to increase crop productivity through improved nitrogen use efficiency; reduced soil emissions, including of nitrous oxide; and increased nonbiochar soil carbon due to improved plant response.

Potential Sources of Leakage and Risks to Emission Reduction
From a climate change perspective the risks related to biochar lie primarily in the negative feedbacks that may occur directly or indirectly during biochar production and application. Such risks include emissions of methane and nitrous oxide during inefficient pyrolysis and degradation of soil organic matter after biochar application on unsuitable soils. Another risk is indirect land use change if biochar use leads to changes in demand for certain types of biomass (a risk that can be minimized if only true wastes are used which is what this report argues for). Also, char dust and other small particulate matter arising from biochar production can become airborne with uncertain global warming impacts, but more definitive adverse impacts on human health. The permanence and long-term stability of the biochar carbon store is another issue, and premature decomposition can be a risk. However, many of these risks can be avoided through the application of appropriate standards and safeguards throughout the biochar production chain, which are currently being developed by a constantly growing community of practice supported by science.

Interaction with Short-Lived Climate Pollutants
Recent science indicates that early and sustained action to reduce black carbon (BC) and methane could reduce near-term warming by up to 0.5°C (UNEP/WMO 2011). Although their lifetime is shorter than traditional greenhouse gases (GHGs) these two pollutants have the ability to warm the atmosphere and surface (particularly when they land on snow or ice causing glacier and ice melt). To the extent that biochar production and use can have an impact on either of these pollutants, it can create another pathway for affecting climate and development. For example, a recent World Bank report (*On Thin Ice: How Cutting Pollution Can Slow Warming and Save Lives*, 2013) shows that biomass cookstoves are significant sources of black carbon with a global impact on the climate and also on human health. Hence efficient biochar production could be an opportunity to reduce short-term warming form BC emissions (depending on the ratio of BC to other copollutants, which may offset some of the BC warming), and contributing to human development by saving lives. Depending on the source of biomass used to produce biochar, methane emissions may also be offset relative to decomposition in open dumps or in fields. Lifecycle emission analysis could compare BC and methane emissions relative to the non-biochar alternative in order to determine net climate benefits.

Life-Cycle Analysis of Climate Impacts
In addition to the carbon stored in the biochar itself a number of important processes can enhance, or reduce, its climate change mitigation potential. These feedback effects need to be accounted for in order to assess the actual net benefits that specific biochar systems can deliver to mitigate climate change. The overall net climate impact of biochar can only be assessed through a full life-cycle assessment (LCA) that takes into consideration the indirect effects listed above,

as well as other secondary sources of GHG emissions (for example transport, energy needed to start the pyrolysis).The LCA is a static snapshot of a situation at a certain point in time. A timeframe can be included that is specific to the product or system analyzed and that represents the impacts at the end of the period. If a series of these calculations are done with different timeframes the development of the emissions balance can be studied over time. More information on LCA, including case studies, is found below.

Potential Global Climate Change Impacts of Biochar at Scale
If biochar were brought to scale what could be the size of its contribution to global mitigation targets? If biochar has the potential to be important globally each project must still be assessed for sustainability and net climate impact on an individual basis, and assess how any wider-scale implementation would impact the global and regional social, biological, and economic systems discussed in this report. It is still too early to estimate what the realistic potential is given all existing economic and social constraints, barriers in technology development, and competing uses in a future bioeconomy. The great advantage of biochar is that it is one of the few GHG reduction strategies that can actually withdraw carbon dioxide from the atmosphere. Over long timescales, however, its main challenge lies in the very fact that biochar will eventually decompose, albeit more slowly than the biomass that it was produced from.

Social Impacts
When designing biochar systems, it is important to take into account that they can affect energy, health, economics, and food security, including in the following ways:

Health and Labor Impacts
Possible impacts can occur throughout the production process. Use of improved cookstoves in biochar systems can increase fuel efficiency and widen the variety of feedstocks. This can reduce pressure on wooded ecosystems and decrease the burden (mainly on women) of fuel gathering. Improved cookstoves can also reduce indoor air pollution providing significant health benefits throughout the developing world especially for women and children. In addition, if biochar improves crop yields or crop resilience the institution of biochar systems could help buffer practitioners against crop shortages and hunger. On the other hand, in the process of producing, storing, transporting, and applying biochar there are potential risks to human health. These need to be addressed by designing appropriate biochar and pyrolysis systems. These risks can include potential emission of toxins and inhalation of dust and small particulate matter with consequences for respiratory health.

Access to Energy through Biomass
The provision of energy through biochar projects could bring benefits at many scales particularly for energy-constrained developing countries. Energy could potentially be used for such critical functions as refrigeration of vaccines, water

pumping, and lighting after sunset. The long-term success of such initiatives depends on adequate technical support, local "ownership" of biomass energy strategies, and conducive regulatory frameworks.

Competing Uses of Biomass

The availability of biomass is a key part of what defines the potential scope of biochar projects. Precisely which categories of biomass could be most appropriate for a biochar system are highly location and system specific. Pyrolysis systems that operate at different scales require different amounts and sources of biomass. Matters of biomass availability, affordability, and alternative uses need to be taken into account. Potential sources and issues arising from their use include:

Purpose-Grown Energy Crops

A number of issues relate to the use of dedicated energy crops for biochar production. This includes the diversion of food crops for fuel, diversion of arable land from food crops, direct and indirect land use change, and whether or not energy crops could truly be constrained to degraded or marginal lands that are unsuitable for food crops.

Biomass Residues

Any time soil is deprived of biomass that would have otherwise decayed in situ and hence protected and enriched the soil there is the potential for loss of nutrients. The costs and benefits must therefore be weighed of leaving biomass in situ versus using it to produce biochar that is then added to the soil. The use of process residues and waste residues, which may normally be sent to the landfill may not result in losses to soils, but each scenario would need to be evaluated independently to determine what the standard or baseline practice is. Similar trade-offs apply when weighing the benefits of using biomass for animal fodder or biochar production. Other competing uses of biomass include bioenergy, building materials, and fuel. As previously mentioned benefits are most assured when use is made of "true waste," especially where the waste in question is simply landfilled providing neither climate nor soil nutrition benefits.

Survey and Typology of Biochar Systems

The use of biochar is a relatively new technique in modern agriculture building on a body of research that is mostly less than a decade old. Dedicated institutional capacity to research and develop biochar applications is only now beginning to emerge. A survey was undertaken to obtain an overview of the status of biochar projects globally, particularly in developing countries. The survey had four purposes: (a) to provide a snapshot of the types of biochar projects that currently exist in developing countries; (b) to gather information about constraints and opportunities in these biochar systems; (c) to develop a typology based on the survey results; and (d) to select a few projects to study the life cycle of GHG emission reductions. The survey was conducted by the International

Biochar Initiative (IBI) who received 154 completed surveys from 41 countries. Most projects were still in their early stages, and only 12 were identified as having adequate data for an LCA. Those projects were sent a follow-up questionnaire to collect additional data.

Some key findings of the initial survey were as follows:

- **Biochar production technology.** Choice of technology was closely tied to available feedstocks. Many respondents (32 percent) were making biochar in traditional charcoal pits, while others had moved on to cleaner, more efficient technologies such as batch retort kilns (41 percent) and continuous kilns. The most common stove design was the top-lit updraft (TLUD) gasifier.
- **Energy use.** Only 51 percent of projects indicated that they were or would be capturing useful energy released during biochar production. Of these, cooking was the largest single energy use. Relatively few projects set their sights on generating electricity.
- **Feedstock choice.** A very wide range of feedstocks was reported using all types of biomass including rice residues, wood residues, manure, weeds, and slash-and-char.
- **Project scale.** The majority of biochar production systems were identified as small, tending toward household- and farm-scale systems.

The survey data was then used to construct a typology of biochar systems categorizing projects based on production technology, energy recovery type, feedstock choice, and scale. Cooking energy, as expected, dominated at the household scale, and projects generating electricity occurring at larger scales. The farm scale was most likely to produce biochar without energy capture indicating that the system may be driven primarily by the agronomic benefits of biochar. As the technique of developing biochar system typologies is refined it will provide a useful methodology for assessing and developing biochar production technologies that are suitable to the scale and energy needs of users, and enable due consideration to be given to all of the other social, ecological, and economic components of the agricultural and energy systems in which biochar may play a role.

Life-Cycle Assessment of Existing Biochar Systems

Life-cycle assessment (LCA) is an ISO 14040 normed methodology to evaluate the environmental flows associated with a product, process, or activity throughout its full life. This involves quantifying energy, resources used, and emissions created. Because of its whole system approach, incorporation of economic costs and transparent methodology LCAs are an appropriate method of analysis for estimating the global warming and net economic impacts of biochar systems. The four main elements to the LCA methodology are (a) goal and scope definition, including system boundaries; (b) inventory analysis, including input and output flows of the system; (c) impact assessment, including environmental consequences and climate change impacts; and (d) interpretation, including contribution analysis, sensitivity analysis, and data quality assessment. The contribution analysis calculates the relative impact of different life-cycle stages (for example, feedstock production, transportation, pyrolysis) to identify those stages with the most impact, and to identify opportunities to reduce GHG emissions and costs of the system. The sensitivity analysis determines how the results may be affected by uncertainty or variability in the data and between systems.

Of the 154 projects that responded to the survey, 25 were selected on the basis of project dataset completeness. From that group, three top projects were selected as case studies for an LCA according to six criteria: degree of integration into local economy, variation in biochar production technology and scale, data availability and quality, replicability in a broad range of contexts, impact on GHG reduction (climate change mitigation effect, expressed as carbon dioxide equivalent or CO_2e), and geographic location. The following were the three projects:

- Kenya: household-scale pyrolysis cookstove producing biochar in addition to providing heat for cooking for subsistence farmers;
- Vietnam: biochar production as a by-product of small-scale rice wafer production from rice husk feedstock;
- Senegal: larger-scale continuous process pyrolysis unit producing biochar used by local onion farmers.

Each project was analyzed using the LCA methodology described earlier. The findings of the analyses are summarized in boxes ES.1, ES.2, and ES.3.

Box ES.1 Summary of Kenya Case Study

The Kenya household pyrolysis cookstove system with biochar returned to soil has the potential for climate change mitigation through carbon sequestration and GHG emission reductions, while also being economically viable for the smallholder farmer. Using the pyrolysis cookstove for household cooking and producing biochar, the net GHG reductions are –1.8 tonnes CO_2e per tonne of dry pyrolysis (secondary) feedstock. The net economic balance is –$1 per tonne of dry matter where no monetary value is assigned to surplus grain, labor savings, or potential carbon credits. If the 62 hours saved in fuelwood collection are monetized, the net return would be $8 per year. If carbon credits were received (at $19 per tonne of CO_2e), the household could return from $27 to $50 per tonne of dry matter, depending on whether avoided emissions are valued the same as carbon sequestered directly in the biochar. If the surplus maize is valued, the net return would be $1 per cropping season of biochar's effectiveness, or $245 for a biochar effectiveness of 50 years. Synergistic benefits of the pyrolysis cookstove project may include decreased indoor air pollution and related illnesses, reduced fuelwood consumption and thereby decreased deforestation pressures, and improved long-term soil fertility. More detailed emissions data for BC and organic carbon are needed to determine if this method results in additional climate benefits through BC reduction.

box continues next page

Box ES.1 Summary of Kenya Case Study (*continued*)

The contribution analysis reveals that the net GHG balance is largely driven by the avoided emissions from the traditional three-stone fire and the sustainability of the primary and secondary cookstove feedstock. If the feedstock is unsustainably harvested and does not regrow, then emissions during cooking would not be offset by the biomass regrowth. Meanwhile, the amount of stable carbon in the biochar contributes 29 percent of the net GHG reductions. Soil nitrous oxide emissions can also play an important role in the GHG balance, thus more data are required to quantify this effect.

The sensitivity analysis reveals that the net GHG balance is relatively insensitive to changes in methane emissions during traditional or pyrolysis cooking, the crop yield response with biochar additions, and the duration of biochar's agronomic effect, within the realistic range tested for these parameters. The surplus maize is most sensitive to the crop yield response and the biochar application rate. Meanwhile, the net revenues are very sensitive to the crop yield response and the price of maize.

This Kenya pyrolysis cookstove system presents itself as a low-risk biochar project with climate change adaptation, economic, and public health benefits for the smallholder farmers implementing these stoves and utilizing the biochar.

Box ES.2 Summary of Vietnam Case Study

The Vietnam rice husk biochar system has potential for climate change adaptation and mitigation through carbon sequestration and GHG emission reductions, while also being economically viable for the smallholder farmer. The net GHG reductions are −0.5 tonnes of CO_2e per tonne of dry rice husk feedstock. The net economic balance is +$948 per tonne of dry matter over the 50 years of biochar's agronomic effect, or +$7 for a one-crop effect. In this project, carbon credits are not received, but the farmer could potentially return an additional $15 per tonne of feedstock with emissions trading at a price of $19 per tonne of CO_2e.

Assuming that rice husk remains a sustainable feedstock, the contribution analysis reveals that the net GHG balance is largely (81 percent) driven by the stable carbon in the biochar. Reduced fertilizer needs and soil carbon accumulation play lesser roles in the GHG balance. Soil nitrous oxide emissions could also play an important role in the GHG balance, thus more data are required to quantify this effect.

The sensitivity analysis reveals that the net GHG balance is relatively insensitive to changes in methane emissions during rice wafer cooking and avoided rice husk burning, the crop yield response with biochar additions, and the duration of biochar's agronomic effect, within the realistic range tested for these parameters. The net revenues are very sensitive to the crop yield response and the duration of biochar's agronomic effect, while the relative impact of the transportation distance of the biochar is less.

This Vietnam rice husk biochar system presents itself as a low-risk biochar project with climate change adaptation and economic benefits for the smallholder farmers incorporating biochar into their practices.

Box ES.3 Summary of Senegal Case Study

The Senegal biochar system has potential for climate change adaptation and mitigation through carbon sequestration and GHG emission reductions, while also being economically viable for the smallholder farmer. The net GHG reductions are –0.4 tonnes of CO_2e per tonne of feedstock. The net economic balance is +$6,696 per tonne of dry matter over the 50 years of biochar's agronomic effect, or +$24 for a one-crop effect. In this project, carbon credits are not received, but the farmer could potentially return an additional $8 per tonne of feedstock with emissions trading at a price of $19 per tonne of CO_2e.

Assuming that the feedstock remains a sustainable source, the contribution analysis reveals that the net GHG balance is largely (87 percent) driven by the stable carbon in the biochar. Soil nitrous oxide emissions could also play an important role in the GHG balance, thus more data are required to quantify this effect.

The sensitivity analysis reveals that the net GHG balance is relatively insensitive to changes in methane emissions during pyrolysis and avoided rice husk decay, the crop yield response with biochar additions, and the duration of biochar's agronomic effect, within the realistic range tested for these parameters. Soil carbon accumulation and transportation of the pyrolysis unit also play small roles in the GHG balance. The net revenues are extremely sensitive to the crop yield response, the duration of biochar's agronomic effect, and the price of onions, while the relative impact of the biochar transportation distance, the biochar price, and the pyrolysis unit production time is less.

This Senegal biochar system presents itself as a low-risk biochar project with climate change adaptation and potentially high economic benefits for smallholder farmers incorporating biochar into their practices.

Case Study Comparison and Conclusions

The main findings emanating from a comparison of the case studies are as follows:

- The net GHG balance of the three studies ranges from –0.4 to –1.8 tonnes of CO_2e per tonne of dry matter. The Kenya cookstove project has the highest amounts of GHG reductions due to the avoided emissions from traditional cooking.
- The Vietnam case study in particular highlights the role of reduced agricultural inputs, specifically fertilizers, in reducing GHG emissions.
- All systems analyzed demonstrate that the emissions from biochar production (whether cookstove or village-scale unit), transportation, and stove or kiln construction are minimal compared to the net balance of the system.
- The most important result regarding the economics is that each project has a very short payback period—within one year of when surplus crops are monetized.
- The yield of the crops to which the biochar is applied plays the largest role in determining the economic balance, implying that the farmer's choice of crops can be as important as the type of soil to which the biochar is applied.

- Another important factor in the economic balance is the capital and operating costs of the biochar production technology. In the case of cookstoves, the capital cost over the lifetime of the stove is small and the operating costs are minimal. In contrast the Senegal village-scale pyrolysis unit has significant costs, which are only offset by the large revenues from surplus crop sales.
- The price the farmer receives for the surplus crop is also important for determining the economic balance, although less so than the crop yield response to biochar.
- The duration of biochar's agronomic effect plays a significant role in the economics of biochar systems in developing countries.
- The biochar application rate is also critical; determining the minimum biochar application rate that still achieves the desired agronomic response will enable farmers to make best use of limited biomass and economic resources.

Biochar projects in developing countries have the potential to reduce GHG emissions and be economically viable, as demonstrated by the life-cycle assessment case studies in Kenya, Vietnam, and Senegal. Ensuring the sustainability of the feedstock for biochar production is the first and most important step in achieving GHG reductions. With the feedstock sustainability in line, the recalcitrant carbon in the biochar is the largest source of direct carbon sequestration by removing atmospheric carbon dioxide and stabilizing it in the biochar. Avoided emissions from traditional biomass management practices, such as traditional cooking, can also play an important role when the feedstock is derived from a nonrenewable source. Avoided rice husk burning or decay is less influential because of the renewability of the resource. Emissions during pyrolysis (biochar production) have only a small impact on the net GHG balance for these systems. Meanwhile, the economics of these projects is largely dependent on the effectiveness of biochar to address soil fertility constraints, the duration of biochar's agronomic effect, the biochar application rate, and the value of the crops to which biochar is applied. Research and development efforts should focus on creating knowledge and understanding of these critical and interdependent parameters. This will enable biochar projects in developing countries to be implemented with the highest probability for improving the livelihoods of smallholder farmers while also mitigating and adapting to climate change.

Aspects of Technology Adoption

Economics of Biochar

The results of the LCA case studies and other biochar analyses demonstrate that biochar systems can offer potential GHG reductions in the range of 0.4–2.0 tonnes of CO_2e per tonne of dry feedstock. However, the implementation and sustainability of all biochar systems, particularly those in developing countries where start-up capital and other funds are limited, are dependent on the economics of these projects. If the economics of the projects are favorable to smallholder farmers, the adoption of biochar systems in smallholder agricultural systems will be facilitated and GHG emissions will be further reduced.

The economics of biochar systems in developing countries are dependent on multiple factors that are specific to the project. Some factors are easily quantified, such as: the cost of the feedstock, the capital and operating costs of the stove or kiln, transportation, the price of biochar and surplus crop yields, and the savings from reduced agricultural inputs. Others are less easily quantified, such as: decreased labor, improved air quality, and increased household food. Furthermore, additional benefits may potentially arise due to decreased deforestation pressures, improved water use efficiency, and climate change adaptation. Finally, although none of the biochar projects receive carbon credits, this contribution may increase revenues as projects progress. Overall, the economics of biochar projects analyzed in the case studies are largely determined by the price farmers receive (or lack thereof) for surplus crops due to biochar additions to the soil.

Engagement with Carbon Markets

A number of basic guidelines are required for a successful biochar project to reduce GHG emissions and to generate emission reductions, as follows:

- **Additionality and baselines.** Establishing the baseline scenario is critical to demonstrating "additionality," which implies that a project would not have taken place under a business as usual scenario and without the incentive provided by the price of carbon.
- **Permanence.** Establishing the fraction of biochar that will be relatively stable is critical to maintaining carbon storage.
- **Leakage and system drivers.** Leakage occurs when emission reductions within a project boundary result in increased emissions elsewhere. For example black carbon emissions to the atmosphere, or "rebound effects," that occur when a more efficient stove encourages more cooking and hence more fuel use than otherwise predicted.
- **Measurement, reporting, and verification.** In order to understand the climate effects of a project, it is essential to be able to quantify and verify its impact. The challenge lies in measuring the impact of the biochar, either directly or indirectly.

There are currently no approved biochar carbon finance methodologies—neither under the Clean Development Mechanism (CDM) nor under any of the voluntary carbon market regimes.

Sociocultural Barriers to Adoption

No matter what the technical potential benefits of biochar are, their realization depends on whether and how people implement biochar systems. This in turn requires a highly location-specific understanding of people and their needs, values, and expectations. To gain a better understanding of sociocultural factors, including barriers, a second survey was drafted and sent to the respondents to the initial survey (described above), receiving 48 responses. Barriers identified included:

• Lack of awareness of biochar and a need for education and demonstration projects to show farmers that making and using biochar would be worth their time;
• Labor barriers, for example extra labor required to operate slash-and-char systems and gather dispersed feedstocks;
• Restricted availability of biochar production technologies;
• Environmental concerns related to changing resource use patterns.

This survey also investigated the prevalence of biochar as a traditional farming practice and found that in many regions, some form of biochar application was a traditional practice that had been swept away by the advent of chemical fertilizers and other twentieth-century methods. Many felt that the existence of the traditional practice made their job of communicating the benefits of biochar much easier.

Regarding the perceived benefits of biochar systems, respondents cited soil improvement, increased crop yields, decreased fertilizer use, improved water use efficiency, clean cookstoves, income benefits, and environmental hygiene.

Respondents were split regarding the importance of carbon offset payments to project viability. Almost exactly half said that carbon payments would be nice but that they were not counting on them. The one remaining quarter said their project could not do without carbon payments and the other quarter replied they were not going to pursue them at all.

Potential Future Involvement of Development Institutions, Including the World Bank

The summary above has demonstrated the wide-ranging potential of biochar systems to contribute to the new paradigm of green growth and development, allied to climate resilience. At the same time, biochar is a relatively new science and many uncertainties exist requiring further research and analysis. Life-cycle assessments are needed, which includes, not just technical and agricultural matters, but also the sociocultural aspects of biochar systems. Given the location-specific nature of biochar systems, a challenge lies in conducting applied long-term research under real-world, developing-country conditions, particularly at scale of implementation.

Institutions like the World Bank, particularly through its technical advisory and convening services, could help to forge effective alliances between the research community and development practitioners on the ground. The Global Inventory of Long-Term Soil-Ecosystem Experiments, established by the Duke's Nicholas School of the Environment and Earth Sciences, is a good example of how such an applied scientific approach could work. As this body of experiences expands it will be possible to refine the criteria of desirable biochar interventions, and establish biochar sustainability standards, which could then serve as a basis for policy regulation or certification schemes.

As an active partner in the Climate and Clean Air Coalition (CCAC) to Reduce Short-lived Climate Pollutants (SLCPs), the World Bank is committed to supporting new and innovative ways to reduce the emissions and impacts of BC and methane. To the extent that biochar production and use result in net reductions of these SLCPs there are significant opportunities to expand biochar activities throughout the CCAC network of partners.

Furthermore, development institutions such as the World Bank could engage in knowledge- and technology-oriented services, as well as financing services for biochar projects or programs. These could include, but are not limited to, development of carbon finance-related methodologies for different biochar systems. The World Bank's carbon funds, and other programs such as those administered by the Global Environment Facility, have a potential role to play in that regard. Given the reluctance of the private sector to engage in nascent, unproven technologies, institutions such as the World Bank and particularly the private sector arm of the World Bank, the International Finance Corporation, will be key players in providing financing services for biochar projects in developing countries over the next few years. Publicly funded demonstration, research, and development will need the engagement of bilateral and multilateral development institutions. The feasibility of biochar pilots may be rapidly assessed when considering key questions such as the following:

- Will the biomass be sourced from true waste sources?
- Will the feedstock be sourced from safe materials?
- Will the quantities of biochar required match the availability of suitable feedstock locally?
- Will the pyrolysis system meet certain levels of conversion efficiency and cleanliness?
- Will the appropriate biochar be applied to appropriate soils?
- Will it be practical during monitoring activities to verify carbon storage of biochar through its application to soils?
- Will local farmers likely adopt the technology after the demonstration phase?

Involving the private sector will be crucial in bridging the funding gap that typically constrains the implementation of new technologies with long lead times and considerable research requirements. Innovative financing solutions will be needed. For example, front-loading the potential carbon benefits of biochar systems with the aim of attracting private investors early on. Again, there is potential for the World Bank to play a role in helping to set up the complex structures required. Direct financing of biochar projects, facilitating research, and providing knowledge services are other ways in which the World Bank and other organizations can support development of biochar systems. Given the wide-ranging potential of biochar systems, it is important to build synergies with other projects and programs.

Finally, it is clear that detailed analyses of current projects need to be carried out across their life cycle in order to assess the costs and benefits, particularly with regard to climate, energy production, and food security. The LCA projects described in this report should be followed up over time, and new ones added. A project catalog should be maintained, and data collection should continue for all biochar projects. Lessons learned will be extremely valuable for future projects and programs, and systematic and robust systems should be set up for collecting data that can be of scientific value in demonstrating the implications of biochar systems for crops, climate, and human well-being.

CHAPTER 1

Introduction

Potential of Biochar

Undoubtedly, three of the biggest challenges of the twenty-first century are the need to almost double food production by 2050 (FAO 2009), to adapt and build resilience to a more and more challenging climatic environment, and to simultaneously achieve a substantial reduction in atmospheric greenhouse gas concentrations (IPCC 2007). The surge of interest in climate-smart farming practices with the aim to improve rural livelihoods while mitigating and adapting to climate change has also sparked curiosity in using biochar as a tool to fight climate change while improving soil fertility. Both public and private sector engagements in biochar necessitate a critical investigation of the opportunities and risks for biochar production and integration into managed systems.

Biochar is the carbon-rich organic matter that remains after heating biomass under the minimization of oxygen during a process called "pyrolysis." Its relevance to preventing deforestation, promoting agricultural resilience, and producing renewable energy, particularly in developing countries, makes it an important issue. At the same time, the potential effects on fertility of its application to soils are diverse, and its climate impact is contingent on the design of the system into which it is integrated, making it a complex issue.

There are a number of reasons why biochar systems might be particularly relevant in developing-country contexts. The potential for biochar to improve soil fertility could result in increased yields from previously degraded soils for smallholder farmers. Improved cookstoves that produce biochar as well as heat for cooking could reduce indoor air pollution and fuel gathering. Both these results could also be beneficial to forests, as enhanced food production capacity could potentially decrease the need to clear more forested land for agriculture, and more efficient cookstoves could decrease wood gathering from forests already in decline.

Increases in Research into Biochar

Although biochar has gathered much interest in the past few years, active biochar production and application systems are just beginning to appear, and there are still gaps in our knowledge of biochar systems. While some characteristics are becoming better understood, questions regarding biochar's impacts on soils, climate, and rural societies remain. Dedicated research started in the late 1990s, apart from some notable early research before 1950 (Lehmann and Joseph 2009), but publication has only accelerated in recent years (figure 1.1).

Content and Purpose of Study

This report offers a review of what is known about opportunities and risks of biochar systems. Its aim is to provide a state-of-the-art overview of current knowledge regarding biochar science. In that sense the report also offers a reconciling view on different scientific opinions about biochar providing an overall account that shows the various perspectives of its science and application. This includes soil and agricultural impacts of biochar, climate change impacts, social impacts, and competing uses of biomass. The report aims to contextualize the current scientific knowledge in order to put it at use to address the development—climate change nexus, including social and environmental sustainability. Ultimately, forward-looking suggestions will be derived from the review part of

Figure 1.1 Acceleration of Published Research on Biochar and Charcoal

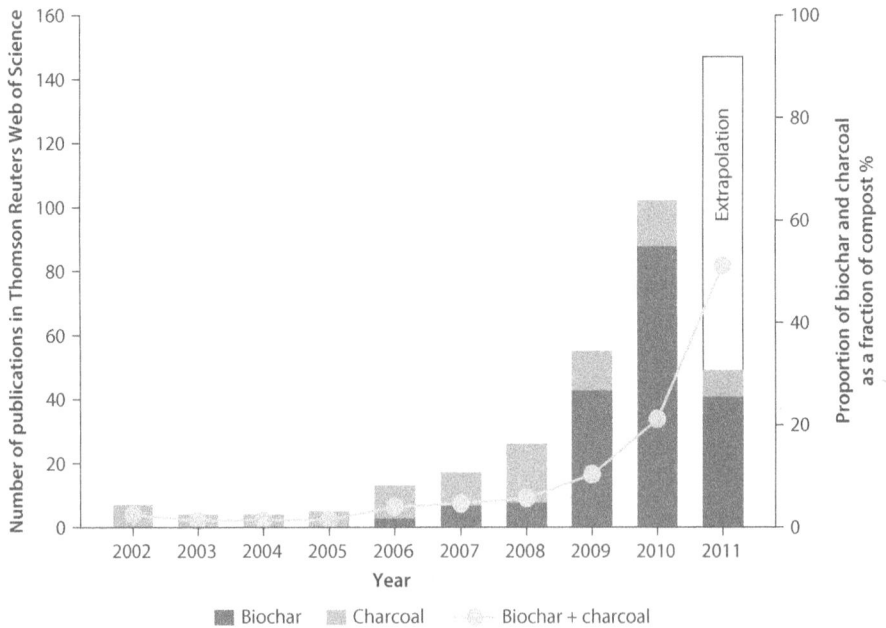

Source: Thomson Reuters Web of Science; only articles with keywords compost, biochar, bio-char, agrichar, agri-char, or charcoal for soil application in the title; data from 2011 are for the first four months and extrapolated in white bar for 12 months.

Biochar Systems for Smallholders in Developing Countries • http://dx.doi.org/10.1596/978-0-8213-9525-7

the report which will show how development institutions, including the World Bank, could support continued research and beneficial application of biochar in development.

The report is organized as follows. The present chapter offers some introductory comments and notes the increasing interest in biochar both from a scientific and from a practitioner's point of view; chapter 2 gives further background on biochar, describing its characteristics and outlining the way in which biochar systems function. Chapter 3 then considers the opportunities and risks of biochar systems. Based on the results of the surveys undertaken (see previous paragraph), chapter 4 presents a typology of biochar systems emerging in practice, particularly in the developing world. New, ISO 14040-based life-cycle assessments of the net climate change impact and the net economic profitability of three biochar systems with data collected from relatively advanced biochar projects were conducted for this report and are presented in chapter 5, providing a novel understanding of the full life-cycle impacts of these known biochar systems. Chapter 6 investigates various aspects of technology adoption, including barriers to implementing promising systems, focusing on economics, carbon market access, and sociocultural barriers. Finally, the status of knowledge regarding biochar systems is interpreted in chapter 7 to determine potential implications for future involvement in biochar research, policy, and project formulation.

Methodology

The method for preparing this report was wide ranging and comprehensive, and included the following elements:

Desk review. A review was undertaken of the current scientific literature pertaining to biochar and issues relevant to its production and application to soils.

Survey of biochar systems. A two-step survey was undertaken to shed light on existing biochar systems, especially those found in developing countries. The initial survey elicited 154 responses, which captured a diversity of biochar projects and biochar systems at different stages of advancement and with different technologies and feedstocks, reflecting differences in the priorities of users. In the second step, an additional survey was drafted and sent to the 154 responders to learn more about the social and cultural barriers to biochar adoption, receiving 48 responses.

Findings from an expert workshop. The report was strengthened and improved through discussions during a workshop held in Washington, DC on May 11, 2011. Comments and advice from the workshop participants helped to inform the analysis of survey data.

Typology. The data derived from the survey were used to develop a preliminary typology of biochar systems, based on production technology, energy use, feedstock choice, and project scale.

Life-cycle assessment. The environmental flows associated with biochar systems were evaluated using a life-cycle assessment, based on the ISO 14040 standard. Three projects—located in Kenya, Vietnam, and Senegal—were selected, based on their integration into the local economy, variation in technology and scale, data availability and quality, replicability, impact on emission reductions, and geographic location. A comparison of the results of the life-cycle assessments was used to draw conclusions that could inform implementation of future biochar projects.

CHAPTER 2

Background on Biochar

Characteristics and Historical Basis of Biochar

Biochar is the solid product remaining after biomass is heated to temperatures typically between 300°C and 700°C under oxygen-deprived conditions, a process known as "pyrolysis." Through pyrolysis, organic materials fundamentally change their chemical composition and are dominated by aromatic carbon forms, in contrast to the original biomass feedstock that mainly contains cellulose, hemicellulose, and lignin. Biochar falls into the spectrum of materials called "black carbon," which includes substances with a range of properties, including slightly charred biomass, charcoal, and soot (Masiello 2004). Black carbon can be produced during the combustion of fossil fuels such as coal or diesel as well as recently living biomass. However, the term "biochar" excludes black carbon derived from fossil fuels or nonbiomass waste, nor would it include materials called "soot" in general usage.

Biochar is best thought of as a system-defined term referring to black carbon that is produced intentionally to manage carbon for climate change mitigation purposes combined with a downstream application to soils for its agricultural effects. This definition of biochar will be used for the purposes of this report. In essence, biochar is distinguished from charcoal in this report by the fact that biochar is produced with the prior intent to be applied to soil, be it as a means of improving soil productivity or carbon storage or both. Also, the array of feedstocks for biochar is much broader than for wood charcoal used as fuel. In principle, biochar can be made from any type of biomass.

Although the term "biochar" has come into common usage only relatively recently, the practice of amending soils with charcoal for fertility management goes back millennia. New instances of traditional practice using biochar are still being discovered around the world. While there is increasing interest in researching new evidence for these ancient practices, for instance in the search for African dark earths (Fairhead and Leach 2009), the currently best-known examples include the ancient practice of adding rice husk charcoal to agricultural soils in Asia (Ogawa and Okimori 2010) and the development of the Amazonian soils known as *terra preta*, or "dark earths" (figure 2.1). Terra preta soils are rich in organic matter and highly fertile compared to the adjacent native soils (Lehmann et al. 2003). Terra preta soils also stand out for their capacity to store carbon, with as much as three times the amount of soil organic carbon compared to surrounding soils (Glaser et al. 2001). They are thought to have been created through soil management by pre-Colombian populations who amended the soils with, among other components, charcoal, giving the soils their characteristic black color. While terra preta can help us understand the effects of biochar in soils, biochars can have a relatively wide range of characteristics. Terra preta soils are the product of many different influences in addition to biochar, and not all its properties are a result of biochar additions.

Some of the most remarkable and currently best-understood properties of biochars include their effects on soil nutrient dynamics and the high stability of the carbon of which they are composed. These properties are considered in more detail in chapter 3.

Because biochar can be produced from almost as many types of feedstock as there are types of biomass, can be created over a wide range of temperatures, and could be applied to a diversity of soil types, it is inappropriate to speak about "biochar" as though it is one homogenous material. Rather, it is important to understand how different production conditions can result in different types of biochars, and how these chars interact with different types of soils. This under-standing is an essential component to the design of any successful biochar system. Ongoing efforts are being led by the International Biochar Initiative (IBI) to develop a characterization standard for biochars (IBI 2012).

Figure 2.1 Terra Preta Soil Pit near Manaus, Brazil, Showing Thick, Dark, Carbon-Enriched Top Layer

Source: Photograph courtesy of T. Whitman, 2010.

Biochar Systems

In order to understand the net impact of biochar, it is essential to understand the entire biochar system (as exemplified by the life cycle assessments investigated in chapter 5). Depending on where the system boundaries are drawn, different elements may be included, but critical to every biochar system are: (a) the source of biomass, (b) the means of biochar production, and (c) whether and how it is applied to soil (figure 2.2).

Biomass Source

Biochar can be produced from almost any type of biomass, including agricultural wastes, rice husks, bagasse, paper products, animal manures, and even urban green waste (Lehmann and Joseph 2009). The biomass used as feedstock is an influential factor for the type of biochar produced and its properties. For example, rice husk-derived biochars have been found to have relatively high ash contents compared to corn cob-derived biochars (Raveendran, Ganesh, and Khilar 1995). The feedstock choice has implications for the ultimate impact of the biochar system. Because many different biomass types can be used for pyrolysis, care must be taken in selecting a biomass feedstock source. Critical factors include its initial moisture content, distance required to transport biomass to the pyrolysis site, whether the biomass source is rural, urban, or industrial, and whether it is

Figure 2.2 Biochar as a System-Defined Concept

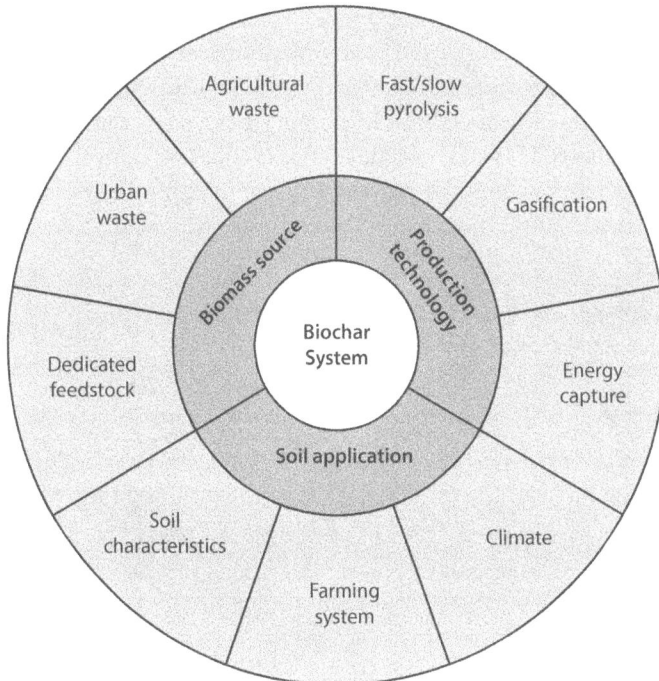

Source: World Bank.

purpose grown or a co- or by-product of another process (Lehmann and Joseph 2009). Focusing on "true wastes" for feedstocks could help to maintain a low-impact system (Whitman, Scholz, and Lehmann 2010). True wastes are materials whose use as feedstock will not result in disruptions in local carbon and nutrient cycling, as when crop residues are removed from fields, or the creation of drivers for other negative land use impacts, such as clearing forest for devoted biofuel crops. The concept of true wastes is tied directly to competing uses of biomass, discussed in chapter 3.

Production Technology

Biochar production can be performed at a very wide range of scales, from the level of a household cookstove in Kenya that produces biochar as well as heat for cooking (Whitman et al. 2011), to the level of an industrial pyrolysis plant that generates both bioenergy and biochar (Gaunt and Cowie 2009; Hammond et al. 2011; Roberts et al. 2010). A system's size and the needs that drive it—whether it is producing energy for cooking, heating, or broader energy needs—are major factors for determining the final impact of such systems.

Biochar-producing systems are usually classified as gasifiers or pyrolyzers and produce in essence three different products that depend on the technology used, namely biochar (solid), syngas (gaseous), and bio-oil (liquid by-product).

Pyrolysis systems use kilns or retorts, and exclude oxygen while allowing the pyrolysis gases, or "syngas,"[1] to escape and be captured for combustion. Pyrolysis systems are classified as slow, fast, and flash pyrolysis, with fast pyrolysis producing more oils and liquids while slow pyrolysis tends to produce more syngas, and flash pyrolysis generating mostly biochar (table 2.1). Gasification systems are usually designed with gas production as the central focus rather than biochar or oil production, and generally produce less biochar than pyrolysis. This is because some oxygen is intentionally introduced in the system; the less oxygen present, the more biochar is produced. However, gasifier systems could be optimized for biochar production, with as much as 30 percent biochar yield predicted to be possible in gasifiers (Brown 2009). For example, rice husk gasifiers are a special

Table 2.1 Typical Product Yields (Dry Basis) for Different Types of Thermochemical Conversion Processes That Generate Carbonaceous Residues

Type	Temperature	Time	Pressure	Liquid (bio-oil) (%)	Solid (biochar) (%)	Gas (syngas) (%)
Hydrothermal carbonization	180–250°C	1–12 hours	Yes	5–20	50–80	2–5
Slow pyrolysis	300–600°C	5–30 min to days	No	30	30	40
Flash pyrolysis	350–650°C	5–30 min	1–3 MPa	8	40	52
Fast pyrolysis	400–550°C	~ 1 sec	No	75	12	13
Gasification	600–900°C	~ 10–20 sec	No	5	10	85

Sources: Antal, Mochidzuki, and Paredes 2003; Bridgwater 2007; Libra et al. 2011; Titirici, Thomas, and Antonietti 2007.

case where the high silica content of rice husks prevents the complete combustion of carbon. As a result, rice husk ash may contain as much as 38 percent carbon (Karve et al. 2011), and is often described not as ash but as "carbonized rice husk." Liquefaction or hydrothermal carbonization is a technology that uses even lower temperatures than pyrolysis, but operates with biomass in liquids under high pressure. Their properties are distinct from biochars, and are not further examined in this report. In the absence of technology access constraints, the preferred technology will depend on users' preferences in the arbitration between energy and biochar production.

Soil Application of Biochar and Its Alternatives

The end use of biochar involves its ultimate application to soil. However, the solid residue from pyrolysis or gasification can also be used for energy through complete combustion, treated as a waste product, or used for a variety of medicinal, industrial, or domestic processes. The optimal end use option depends on the values and objectives of the biochar producer, such as climate change impact, energy production, soil fertility, and economics (as discussed in chapter 3).

Gaunt and Lehmann (2008) found that avoided greenhouse gas (GHG) emissions in a modeled industrial pyrolysis system are between 2 and 5 times greater if biochar is applied to agricultural land as opposed to being solely burned as a fossil fuel. If climate change mitigation was the producer's objective, then biochar application to soil would be the clear choice for its end use. The global model developed by Woolf et al. (2010) to estimate the total contribution of sustainable biochar to climate change mitigation expands this finding. It concludes that biochar has greater mitigation potential applied to soil than as an energy source only if it generates soil productivity benefits or reduces emissions of non-carbon dioxide GHGs from soil. In high-energy-demand scenarios, such as in households where there is a shortage of fuel, combusting biochar for energy could conceivably be chosen over applying it to soils. If biochar were consumed for energy, rather than applied to soils or reserved as a store of carbon, then no agronomic effects of biochar would occur, nor would there be any long-term storage of carbon. Still, this flexibility could actually be considered an advantage of biochar systems—giving the user greater economic flexibility to utilize the biochar produced for either fuel or soil building, or both, depending on the need at the time.

If biochar is neither used for energy, nor applied to soils, it has been suggested that it may simply be stored in geologic reservoirs, such as landfills, abandoned mines, or at sea, as described by Seifritz (1993) and Lee et al. (2010). But, in this scenario, the possible agronomic effects from biochar application to soils would certainly not occur. This would completely remove a potentially important driver for the system—any effects of biochar application on soil fertility—and would result in the withdrawal not only of carbon from the soil system, but also of all other nutrients that may be stored in the biochar. It should be noted, however, that biochar conversion in some systems also reduces nutrient return to soil, mainly the amounts of nitrogen and sulfur, compared to leaving crop residues in the field (chapter 3).

Thus, for the purposes of this report, the focus is on integrated bioenergy-biochar-soil application systems, and the two scenarios described above will not be considered other than as alternative uses.

During the application of biochar to soil, two major aspects to consider are the appropriate selection of biochars for specific soil constraints and the actual biochar application methods. The recognition of appropriate biochar-soil combinations or biochar and specific farming practices is essential for any successful biochar system. As discussed further in chapter 3, because biochars and soils can have diverse properties, it is necessary to understand the characteristics of both in order to achieve an effective pairing. Because biochar is a newer technology in modern farming (than, for example, compost or mineral fertilizer management), relatively little work has been done to develop optimal application strategies for agricultural (or other) soils. Biochar could be applied on its own or mixed with other additions such as manure, compost, or mineral fertilizers. It could be incorporated into the topsoil during plowing; through banding (localized application); by top dressing (applying to soil surface and allowing natural processes to incorporate it) (Verheijen et al. 2010); in planting holes (Blackwell, Riethmuller, and Collins 2009); as seed coating; in planting tubes; or when soils are built for establishing green roofs (Beck, Johnson, and Spolek 2011) or recreational turf.

Note

1. Syngas production from fast and slow pyrolysis depends on the specific system, but is generally composed of a range of gases, including CO, CO_2, H_2, CH_4, and low-molecular-weight hydrocarbons such as C_2–C_6 (Ioannidou et al. 2009).

CHAPTER 3

Opportunities and Risks of Biochar Systems

Introduction

In order to assess the net climate impact and sustainability of a biochar system, it is important to understand the effects of biochar on soil health and productivity, and on greenhouse gas (GHG) emissions. A number of studies have demonstrated that biochar soil management has potential benefits for soil fertility and agricultural productivity, though the range of system variables (including soil and biochar characteristics) makes precise analysis difficult. Similarly, the prospect of climate change mitigation through biochar technologies has attracted widespread attention. Possibilities for emission reductions exist through carbon stabilization, renewable energy applications, and waste diversion. A number of challenges remain to be overcome, however, to ensure that benefits are optimized and leakage of GHGs is avoided. In that regard, efforts should be made to ensure that only "true wastes" are used for biochar production. Undertaking a life-cycle assessment is essential in order to evaluate the net climate impact of biochar systems.

It is vital that any review of the potential benefits of biochar systems extends beyond consideration of the technological and scientific aspects to take account of the social dimensions. While various positive impacts have been identified, including increased fuel efficiency leading to a reduction in indoor air pollution and less time spent gathering fuelwood (with particular benefits for women), potential risks are also present, for example threats to health from inhalation of black carbon and toxins. In addition, challenges exist in ensuring that the benefits from biochar systems are shared equitably, and that nonbeneficial changes in land use do not result from the addition of biochar to the competing demands for biomass.

This chapter aims to present the current state of knowledge on these opportunities and risks of biochar systems, while recognizing that the complexity of the issues surrounding biochar, and the relative recent recognition of its great potential, means that many knowledge gaps exist and a great deal of research is needed to gain a true idea of the benefits it can offer.

Impacts on Soil Health and Agricultural Productivity

The potential effects of biochar on soil fertility and agricultural productivity are widely reported (Blackwell, Riethmuller, and Collins, 2009; Chan and Xu 2009; Lehmann and Rondon 2006; Sohi et al. 2010; Verheijen et al. 2010). As described in chapter 2, terra preta is a striking illustration of the potential for improving the fertility of highly weathered tropical soils. In a meta-analysis of 86 different biochar treatments on a range of soil and crop types, Verheijen et al. (2010) reported a grand mean of 10 percent increase in plant productivity. In addition, a growing body of literature indicates that biochar amendments are in some situations able to increase yields substantially, sometimes by more than 100 percent compared to a fully fertilized control (for example, Steiner et al. 2007; Kimetu et al. 2008; van Zwieten, Kimber, Morris, Chan et al. 2010). The potential agronomic benefits reported by biochar practitioners are not only greater yields, but also better predictability in yields, reduced germination time, extension of the cropping season, and improved resilience to drought (Hayes 2010; Couto 2010; Reinaud 2010). Looking at the properties of some specific biochars, it is reasonable to expect that biochar soil management in some circumstances could even open new possibilities for cropping, producing crops on normally unsuitable sites (Lehmann and Rondon 2006).

Key Factors That Influence Biochar's Impact

The high variability of biochars and soils requires systematic studies across soils to identify important variations between different biochars, soil types, and quantities applied. In addition, the physical structure and chemical properties of biochars evolve over time (Cheng et al. 2006; Cheng, Lehmann, and Engelhard 2008; Nguyen et al. 2008), which makes predicting the long-term effects on soil fertility and agricultural productivity through short-term experimentation more challenging. Studies are often difficult to compare as they use biochars of different origins and production methods, different soil types, and different climatic conditions, and often track only plant growth as a proxy for crop yields. One important problem with short-term experiments is that some improvement in plant productivity is likely attributable to the nutrient-rich ash content of the char, which provides a real but relatively short-lived fertilizing effect. This section will review the mechanisms and risks associated with some of the key factors that influence the impact of biochar on soil health and agricultural productivity, including soil pH, nutrient availability, soil moisture, soil organic matter, and the amount of biochar applied.

Soil pH

Biochar pH values can range from acid to basic (pH from < 3 to > 12) (Cheng et al. 2006; Lehmann 2007). Most of the biochars currently in use have a neutral to basic pH,[1] which is why biochar applied in acidic soils tends to cause a liming effect. In their study, Verheijen et al. (2010) found that the soil pH rose on average from pH 5.3 to pH 6.2 after biochar application. There was a significantly higher gain in plant productivity following biochar addition in acidic

Figure 3.1 Percentage Change in Crop Productivity upon Application of Biochar under Different Scenarios

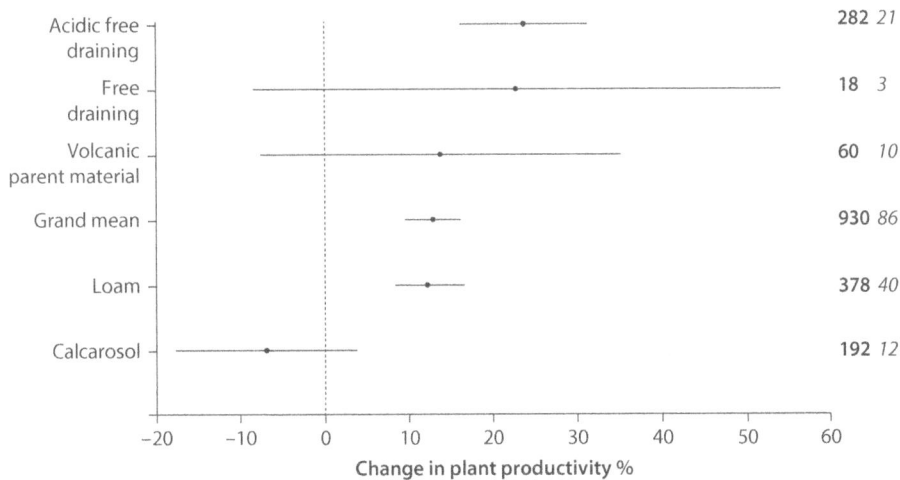

Source: Verheijen et al. 2010.
Note: Figure shows percentage change in crop productivity upon application of biochar at different rates along with varying fertilizer coamendments to a range of different soils. Points show mean and bars show 95 percent confidence intervals. The numbers in the two columns on the right show number of total "replicates" upon which the statistical analysis is based (bold) and the number of "experimental treatments" that have been grouped for each analysis (italics).

free-draining soils than in calcarosols (alkaline soils formed from calcium-rich sediments of variable texture), where the trend is rather negative (figure 3.1). The authors conclude that their results provide evidence to support the increase of plant productivity due to the liming effect of biochar. Biochar thus has the potential to reduce the need for conventional liming operations, cutting back the costs associated with this practice.

However, the effects of biochar pH values vary. In basic soils, biochar application may be a hindrance: van Zwieten, Kimber, Morris, Chan et al. (2010) found that additions of a biochar with a pH of 9.4 to a basic soil with a pH of 7.7 decreased crop yields. In addition, the pH of the biochar itself will change over time in soils. The magnitude of this change in pH varies with type of feedstock used to produce the biochar, as well as the production conditions. It may increase for biochars that contain substantial amounts of base cations, or decrease for biochars made from woody species (Nguyen and Lehmann 2009).

Nutrient Availability
The impact of biochar on nutrients present in the soil and their bioavailability to plants is very important and can occur in many ways (see table 3.1 for a discussion of nitrogen as an example). Biochar is not considered a fertilizer per se but rather a soil conditioner that affects the retention and mobilization of existing nutrients for plant uptake. Biochar can help improve soil functions by reducing nutrient loss below the root zone, increasing the bioavailability of nutrients to plants over time (Shackley and Sohi 2010) and decreasing nutrient leaching

Table 3.1 Possible Biochar Effects on Nitrogen Cycling

Effect category	Possible effects	Implications for nitrogen	References
Biological and nutrient cycling effects	Biochar loses 50% or more of its nitrogen (N) during pyrolysis and has relatively more carbon (C) than N compared to other soil organic matter (a high C:N ratio)	Over the short term, microorganisms trying to decompose the high-C biochar may immobilize ("tie up") N that would have been available to plants	Clough et al. 2004; Lang, Jensen, and Jensen 2005; Novak et al. 2010; Robertson and Groffman 2007; Rondon et al. 2007
	Different microorganisms are active at different pHs	Certain microorganisms responsible for nitrous oxide (N_2O) production are more abundant at lower pH	
	Different microorganisms are active at different oxygen contents, which are affected by how much water is filling soil pore space	Microbially driven nitrogen cycling is affected by aeration—for example, N_2O is produced under anoxic, water-logged conditions	
Chemical properties of biochar	Biochar may retain positively charged ions, depending on its cation exchange capacity (CEC), which may increase over time	CEC could potentially retain nitrogen in the form of ammonium (NH_4^+)—more developed in older biochars	Cheng et al. 2006; Cheng, Lehmann, and Engelhard 2008; Laird et al. 2010
	Biochar is less likely to have anion exchange capacity (AEC)	Changes in nitrate (NO_3^+) retention are negligible	
Chemical changes in the soil	Biochar addition to soil may change pH	A neutral pH will increase N mineralization and availability	Chan and Xu 2009
	Biochar addition to soil may add nutrients directly	Some biochar-N will be present, but a large portion of it is lost during production and most of the N remaining in the biochar is unavailable to plants over short periods of time	
Physical changes in the soil	Biochar may change porosity and water-holding capacity, altering water dynamics	If water movement through soil is decreased, leaching of nitrogen, particularly in the form of NO_3^+, could be reduced	DeLuca et al. 2006; Laird et al. 2010
		Increased adsorption of phenols in forest soils will increase nitrification	

Source: World Bank.

(Dünisch et al. 2007; Laird et al. 2010; Lehmann et al. 2003; Novak et al. 2009; Steiner et al. 2007), likely due to the porosity and internal surface area of the biochar particles. However, the ability to improve nutrient availability may vary depending on the plant species. Woolf et al. (2010) reviewed available field and greenhouse data and found that cereals responded roughly three times more than leguminous species.

Biochar's ability to improve nutrient availability has the potential to reduce reliance on fertilizer and its associated environmental consequences (Verheijen et al. 2010). This is why biochar is of great interest to countries such as Australia that rely on a costly foreign fertilizer supply (CSIRO 2010). Similarly, the application of fertilizing agents jointly with biochar has been observed to increase the positive response to the biochar in various cases (Chan et al. 2007; van Zwieten, Kimber, Downie et al. 2010). One of the techniques in traditional Chinese agriculture may have been the use of rice husk biochar mixed with "night soil" (Beagle 1978). Recent studies have also found that biochar can reduce nitrogen losses during composting of manure (Steiner et al. 2010).

Fresh biochar can also contain nutrients that were present in the original feedstock. Any nutrient that does not volatilize during pyrolysis is conserved, which is the case for most nutrients except nitrogen and sulfur (Lang, Jensen, and Jensen 2005; Shackley and Sohi 2010). When the biochar is applied to the same piece of land that produced the pyrolyzed feedstock, the nutrient cycle remains broadly balanced. Any geographic separation between biochar production (feedstock sourcing) and biochar application will likely create disruptions to the local nutrient cycle and may jeopardize long-term fertility. Similarly, if biochar production uses feedstocks grown remotely from the land receiving the biochar, there is a risk of depriving some areas of biomass and associated nutrients in order to enrich others. Closed-loop systems would avoid this problem.

Intriguingly, using specific wastes, such as urban green wastes, could potentially establish a flow of nutrients from cities back to rural areas, particularly in regions progressively drained of their nutrients as they feed growing cities. Not only could this provide a closed-loop biochar system, it may help solve an important waste management challenge and provide new business opportunities. However, the quality of urban waste must be taken into consideration: biochar could concentrate toxic elements (such as heavy metals), and land contamination is a risk if biochar is produced from waste of unknown composition.

Soil Moisture

Biochar can improve water infiltration and retention in soils through a combination of direct and indirect mechanisms. The direct effect of biochar-induced water retention is related to its porous structure and large surface area. Plant-available water is better retained in soils with intermediate pores (0.2–20 micrometers). Biochar itself holds greater amounts of plant-available water, and the changes to soil after biochar application also improve overall water-holding capacity (Sohi et al. 2010). Additional studies have shown that biochar can improve porosity, infiltration, and water efficiency in a number of soil types (for example, Kammann et al. 2011; Ayodele et al. 2009); however, for some types of soils, such as clay soils with a high water retention capacity, biochar is less beneficial or even a hindrance (Busscher et al. 2010; Martin and Moody 2001; Tryon 1948). It is thus important to characterize the porosity of different biochars over time and target their application. For example, field studies with biochar application to sandy soils with a low water-holding capacity have shown positive results, as this is where improved water retention is most needed.

The indirect, longer-term effects of biochar on soil moisture are also positive, though the processes remain much less understood. According to Glaser, Lehmann, and Zech (2002), terra preta that received biochar several thousand years ago retains 18 percent more water than adjacent soils with low biochar contents. Studies have postulated that biochar may improve the ability to retain plant-available soil moisture and increase infiltration in the long term due to greater amounts of soil organic matter and increased aggregation resulting from microorganisms (Lehmann et al. 2011; Tisdall and Oades 1982).

Soil Organic Matter

Biochar cannot replace the portion of organic matter that supplies the soil biota with energy, and the impact of biochar application on the preexisting soil organic matter is thus fundamental. In tropical regions, where climatic conditions generally result in faster cycling of soil organic matter, the depth of organic carbon accrual found in terra preta sites has raised some expectations on the indirect role that biochar could play in generating and sustaining high levels of soil organic matter. Indeed, plant litter input to terra preta soil was stabilized to a greater extent than litter added to adjacent soils with low biochar contents (Liang et al. 2010).

That said, studies have shown a short-lived "priming" effect: an increased mineralization of existing soil organic matter and disappearance of litter, likely due to direct impacts of biochar application, such as pH changes and nutrient additions (Hamer et al. 2004; Wardle, Nilsson, and Zackrisson 2008). However, greater stabilization of soil organic matter was observed in the long term (Kuzyakov et al. 2009; Zimmerman, Gao, and Ahn 2011). In a two-year field study, Major et al. (2010) found that long-term increases in plant growth likely played a greater role in improving stocks of soil organic matter than direct soil changes. As with moisture-rich clay soils and high-pH basic soils, evidence indicates that biochar added to peat or soils with very high concentrations of organic matter is less likely to increase soil fertility and may even reduce it (Kimetu et al. 2008).

Amount of Biochar Applied

Plant response to biochar also varies with the amount of biochar applied. Another recent meta-analysis of 19 published field and greenhouse trials (Crane-Droesch et al. 2010) found that plant biomass increased on average by 3.8 percent ± 2.4 percent per tonne of biochar applied per hectare at moderate application rates. For example, according to this study 10 tonnes of char per hectare would increase plant biomass by 38 percent on average. With optimum conditions (which remain to be fully characterized, but an example might be drought-susceptible sandy acidic soils paired with a basic biochar), yield increases may be even higher.

A positive correlation between increasing levels of biochar addition and plant response may only persist up to a certain point (Kammann et al. 2011; Lehmann and Rondon 2006; Rondon et al. 2007). At higher application rates, the effect on yield may decrease, and may in some cases become negative (Crane-Droesch et al. 2010; Rondon et al. 2007). The specific biophysical optimum for each soil–crop system remains to be determined. It may vary from a few tonnes per hectare for certain situations, to much larger quantities in other circumstances.

The economic optimum, however, may lie significantly below the biophysical optimum, considering costs of biochar applications and expected revenues from gains in crop yield.

Important agronomic benefits have been observed even with rates of biochar application that are obtainable by subsistence farmers. Lehmann and Rondon (2006) noted improvements in productivity from 20 percent to 220 percent at application rates of 0.4 to 8 tonnes of carbon per hectare on degraded soils in Kenya. Kimetu et al. (2008) report a doubling of maize yields after applying 8 tonnes of biochar per hectare in the central Amazon, and Steiner et al. (2007) found similar increases for rice with 8 tonnes of biochar per hectare. This is particularly interesting as smallholders in developing countries would be expected to produce relatively small quantities of biochar, either due to a lack of unused biomass available locally or due to other constraints specific to smallholders. Typically, maize residues allow application of 0.5–2 tonnes of biochar per hectare per season (using slow pyrolysis) when recognizing the need for a 50 percent crop residue return to soils. Biochar production by smallholders using a pyrolysis cookstove would be limited according to their cooking needs. In practice, smallholders may never produce more than 1 tonne of biochar per year per household (Whitman et al. 2011). In western Kenya, cooking with a pyrolysis stove generates approximately 0.5 tonnes of biochar per hectare per year (Torres et al. 2011).

Toward Optimization of Biochar Use

Poor design of biochar systems runs the same crop failure risk as other nonchemical techniques that aim to improve agricultural productivity. In agroforestry, for instance, the wrong tree–crop combination may increase plant competition for light and water and lead to crop failure. However, unlike agroforestry, which now benefits from decades of on-farm research, biochar use is still hampered by a lack of experience and understanding of the enabling conditions for securing predictable positive effects. Knowledge of optimized biochar–soil–crop combinations is incipient. The use of sound pilot tests before any full-scale, on-the-ground implementation is therefore highly recommended for any biochar activity. The fact that biochar, once incorporated, cannot be removed from the soil calls for a precautionary approach.

More research is needed to explore the duration of biochar's effect on soil properties and thus crop productivity. If the positive biochar effect on the soil is due to improved cation exchange capacity, increased soil organic carbon, or improved soil water-holding capacity, then the effects can be expected to be long term. This has been demonstrated in terra preta soils (Liang et al. 2006). However, if the positive biochar effect is due to improvement in pH or acidity decline or nutrients added with the biochar, then the duration of these positive effects could potentially disappear before the biochar decomposes. An understanding of the soil constraint that the biochar addresses will help estimate the expected duration of the biochar's positive effects (Lehmann 2009).

There is a relatively safe array of biochar applications where risks of crop failure are limited. For example, drought-susceptible sandy or acidic soils seem to be particularly suited for the addition of biochar with basic pH and no harmful contaminants to increase crop yields. Evidence indicates that biochar application to clayey soils, high pH soils (Asai et al. 2009; van Zwieten, Kimber, Morris, Chan et al. 2010), peat, or soils with very high concentrations of organic matter (Kimetu et al. 2008) is less likely to increase soil fertility and may even reduce it. Given the uncertainties that remain around the interactions between biochar and preexisting soil organic matter, it may be useful initially to focus on soils that are poor in organic matter. The abundance of marginal or degraded soils in many parts of the tropics, in particular, means that biochar's potential in agriculture could be large.

Impacts on Climate Change

Biochar is a multifaceted technology and its relevance to climate change mitigation has probably caught as much attention as its potential for agronomic purposes. The biochar material itself is a carbon concentrate of the original biomass feedstock, and its stability in soil thus has implications for carbon sequestration. A number of important processes can enhance or reduce its climate change mitigation potential at different stages of the biochar system, including, for instance, land use changes occurring before biochar application (sourcing of biomass for pyrolysis) and after (plant response). This section reviews the direct and indirect ways biochar prevents and sequesters GHG emissions, the potential risks and sources of leakage, and the importance of viewing the biochar systems from a life-cycle perspective. Finally, the potential for biochar to make a contribution to regional and global mitigation targets is discussed.

Direct and Indirect Sources of Emission Reductions

One of the largest differences between emission reductions from biochar and bioenergy is the former's benefits when added to soil, such as greater plant growth and reduced soil emissions (Roberts et al. 2010; Woolf et al. 2010). Carbon storage and stabilization is probably the most direct and most important quality for climate change mitigation efforts based on biochar. Other relevant processes also come into play indirectly, including the generation of renewable energy, waste diversion, reduced fertilizer manufacturing, reduced soil emissions, and increased nonbiochar soil carbon. A summary of all sources of emission reductions from biochar is provided in table 3.2.

Carbon Stabilization

Biomass typically contains about 50 percent carbon, which is relatively quickly decomposed and reemitted to the atmosphere upon decay in soil. The mean residence time of fresh biomass is in the range of months to years, with longer times for woody biomass and colder climates (Mungai and Motavalli 2006). Biochar retains between 10 percent and 70 percent (on average about 50 percent) of the

Table 3.2 Direct and Indirect Sources of Biochar Emission Reductions

Source	Description	Trade-off	Dominant GHG	Relative importance for total GHG balance[a]
Carbon stabilization	Biochar decomposes more slowly than the biomass from which it was produced, taking into account the initial carbon loss during pyrolysis.	Potential energy production is lost due to retention of biomass fuel as biochar.	Carbon dioxide (CO_2)	50–65% (26–42% in cookstove system)
Renewable energy	The energy produced during pyrolysis can be used to replace energy needs normally filled by fossil fuels or unsustainably harvested biomass.	Minimal trade-offs exist if pyrolysis is an efficient and appropriate energy source for the system.	CO_2 (N_2O and CH_4 in inefficient systems)	20–40% (56–72% in cookstove system)
Waste diversion	If the feedstock biomass would have decayed under low-oxygen conditions, such as in a landfill, then methane (CH_4) and nitrous oxide (N_2O) would be emitted, increasing the climate impact of the baseline scenario.	No CH_4 is captured for bioenergy through digestion.	CH_4, N_2O	0–75%
Reduction in fertilizer manufacturing	If biochar production reduces the need for fertilizers, the energy and emissions associated with their production would be reduced.	If nitrogen immobilization occurs for a limited time after biochar additions due to mineralization of the labile fraction of biochar, nitrogen fertilizer requirements may actually be increased.	N_2O	Not quantified
Reduction in soil emissions	Biochar application to soils may reduce net emissions of CH_4 and N_2O. These effects are currently poorly understood. If biochar reduces nitrogen fertilizer applications, this could directly reduce N_2O emissions.	Biochar applications to soil may increase emissions of CH_4 and N_2O. If nitrogen immobilization occurs for a limited time after biochar additions due to mineralization of the labile fraction of biochar, additional nitrogen fertilizer is needed.	CH_4, N_2O	0–5%
Increased non-BC[b] soil carbon	Biochar application may increase plant growth and the associated inputs to soil carbon through residues and root growth. Biochar may also stabilize soil carbon, reducing decomposition.	Biochar may initially increase mineralization of soil carbon (the "priming effect") through additions of nutrients or labile carbon, or pH increases (transient phenomenon).	CO_2	~ 2% from increased soil organic carbon inputs in cookstove system; carbon stabilization not quantified

Source: World Bank.

a. Estimates of relative importance calculated with data from Gaunt and Cowie 2009; Hammond et al. 2011; Roberts et al. 2010; Woolf et al. 2010 (industrial systems); and Whitman et al. 2011 (cookstove system)

b. BC = black carbon.

carbon present in the original biomass (Lehmann et al. 2003), and slows down the rate of carbon decomposition by one or two orders of magnitude, that is, in the scale of centuries or millennia (Cheng, Lehmann, and Engelhard 2008; Lehmann et al. 2008; Spokas 2010; Zimmerman 2010). The most important issue from a carbon point of view is the *difference* between the decomposition of the biochar and the decomposition of the original feedstock that would have happened in the absence of pyrolysis. The relative decomposition rate between biomass and biochar is affected by in-soil decomposition rate, the biochar's labile to recalcitrant ratio (see below), and the pyrolysis technique used to produce the biochar.

Many factors determine the rate at which decomposition occurs, including chemistry of the biochar and biomass and the soil conditions. Biochar has relatively more bonds that are harder to break apart, such as carbon-carbon double bonds in interconnected aromatic ring structures of variable sizes (Keiluweit et al. 2010; Nguyen et al. 2010), making it slower to decompose than most other forms of organic matter. Decomposition can also occur at very different rates from one environment to the next. Organisms are more active and decompose more organic matter in ideal soil conditions (Cheng, Lehmann, and Engelhard 2008; Nguyen and Lehmann 2009; Nguyen et al. 2010). The presence of mineral matter in soil can also slow decomposition and enhance carbon stabilization (Glaser et al. 2000; Lehmann and Solomon 2010; von Lützow et al. 2006; Nguyen et al. 2008).

All evidence from field and laboratory studies shows that biochar decomposition is initially rapid over timescales of weeks and months, after which the rate of loss decreases (Brodowski 2004; Kuzyakov et al. 2009; Zimmerman 2010). This variation in decomposition times suggests the presence of a labile fraction in the biochar that degrades rapidly, distinct from the recalcitrant fraction that may take millennia to completely disappear.[2] Therefore, the carbon sequestration

Figure 3.2 General Concept of the Carbon Storage Potential of Biochar Based on 1 Tonne (t) of Dry Feedstock (Slow Pyrolysis)

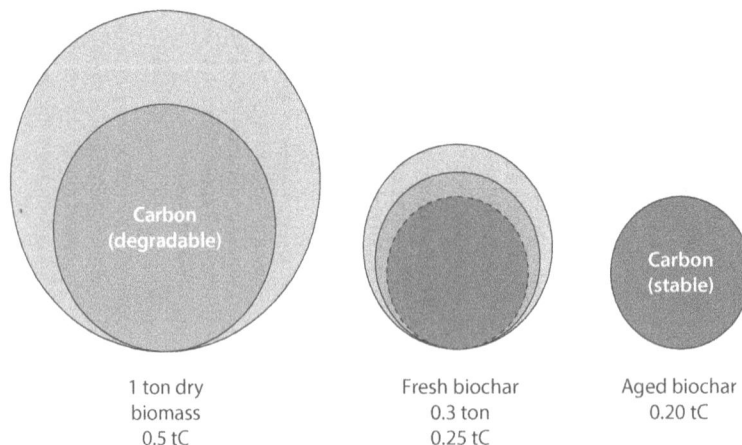

1 ton dry	Fresh biochar	Aged biochar
biomass	0.3 ton	0.20 tC
0.5 tC	0.25 tC	

Source: World Bank.

potential of biochar may be determined by the relative sizes of the recalcitrant and labile fractions. Estimates of these fractions would be expected to vary from one biochar to the next, and would depend on a number of factors, such as initial feedstock, pyrolysis temperature, and pyrolysis duration. Only rudimentary guidance currently exists in the literature about biochar properties that predict recalcitrance. For example, one of the principal bottlenecks for the establishment of a sound biochar protocol for carbon markets is the so-called volatile matter of the oxygen to carbon (O:C) ratios (Spokas 2010; Zimmerman 2010) (see chapter 6 for further discussion). According to model-derived estimates, delivering carbon sequestration over the next few hundred years would require either the labile fraction of the biochar to be below 10 percent or the entire biochar to have a mean residence time[3] of a thousand years or more (figure 3.2).

In addition to the labile to recalcitrant ratio of the biochar, the pyrolysis system used to create the biochar also affects the relative rate of decomposition by impacting the amount of carbon captured in the biochar. Some biochar production techniques may capture less than 40 percent of original carbon due to equipment inefficiency (for example, during initial development stages of biochar stoves), the type of feedstock used (for example, carbon retention through pyrolysis is greater with woody biomass than with manure), or the type of biochar production technology (gasifiers favor a lower biochar to bioenergy production ratio compared to pyrolyzers; see table 2.1). Greater pyrolysis temperatures usually decrease the carbon captured in the biochar (Lehmann 2007), while also increasing the stability of biochars (Zimmerman 2010) and the proportion of stable biochar (Nguyen and Lehmann 2009).

In theory, biochar produced through slow pyrolysis can store for the long term about 1 tonne of "stable" carbon for each 5 tonnes of dry feedstock. As shown in figure 3.2, the estimates used to reach this 1:5 ratio assumes the following: dry biomass is 50 percent carbon, 50 percent of the carbon is lost during biochar production through pyrolysis, and 80 percent of the remaining carbon is relatively recalcitrant. These numbers correspond to biochars produced through slow pyrolysis and would change if pyrolysis conditions change (Spokas 2010; Zimmerman 2010). Hydrothermal carbonization, for instance, can virtually suppress carbon dioxide emissions during pyrolysis, but at the expense of a much less resistant biochar (half-life under 30 years according to Woolf et al. (2010); for example, see Steinbeiss, Gleixner, and Antonietti 2009). The understanding of biochar's potential for carbon storage—particularly given the diversity of feedstocks, environmental conditions, and pyrolysis techniques—is an area that requires further studies and research.

Renewable Energy and Energy Efficiency
In addition to carbon stabilization, biochar systems may also displace business-as-usual emissions to provide indirect sources of emission reductions. A certain amount of GHG emissions are produced during pyrolysis; however, this risk does not feature prominently in biochar literature, as it is assumed that the methane produced during pyrolysis will be combusted and used as a source of renewable

energy (including the energy required to run the pyrolysis unit). Most biochar technologies are tailored for combustion of the syngas by-product of biochar production. Even very small-scale, highly disseminated technologies, such as biochar stoves, can be efficient in using this syngas for heat energy. Less efficient pyrolysis technologies, such as biochar kilns that aim to maximize char production without combined valorization of the volatile energy released, should be carefully scrutinized in this respect.

An important double climatic benefit occurs when not only the potential methane emissions are avoided, but the useful renewable energy is used to offset fossil fuel combustion. This is, however, contingent on both the "renewability" of the particular system design and biomass source, and having a surplus of energy after using what is needed for operating the pyrolysis unit. As the pyrolysis process is exothermic, surplus energy is usually available (Bridgwater 2007; Brown 2009), but it may be compromised by the use of wet feedstocks that require a portion of the energy to be used for drying.

In certain circumstances, the energy produced through pyrolysis may not offset fossil fuel use, but may reduce unsustainable wood harvesting. For instance, the introduction of efficient biochar cookstoves may replace less efficient methods for cooking. However, nonpyrolytic improved combustion cookstoves are also much more efficient than traditional three-stone fires (Johnson et al. 2008; MacCarty et al. 2008), and benchmarking against the best available technology has not been done to a sufficient extent (Whitman et al. 2011). Fuel savings can also occur more indirectly when soil enriched with biochar reduces the need for cultivation and irrigation (for example, reducing fuel used for plowing), but this is extremely context dependent and challenging to quantify.

Waste Diversion

Biochar systems also avoid emissions in situations where biochar is produced from wastes that would normally decompose anaerobically, releasing methane and nitrous oxide in the process. However, it is important to consider what constitutes "true waste." Typically, all biomass that is usually burned without any nutrient or energy capture or left to decay off farm (for example, sawdust, municipal green waste) with no other use can be considered a true waste.

Reduction in Fertilizer Manufacturing

The impact of biochar on nitrous oxide emissions is promising if biochar can indeed increase crop productivity through improved nitrogen use efficiency (Chan et al. 2007; Steiner et al. 2008; van Zwieten, Kimber, Morris, Chan et al. 2010) and help farmers cut back on the use of nitrogen fertilizers. However, it remains to be fully assessed how significant such an effect of biochar application might be as it depends on soil type, climate, local weather conditions, crop, and farmers' preferences. Roberts et al. (2010) report that the nitrous oxide avoidance effect accounts for probably not more than 2–3 percent of the overall GHG abatement potential of biochar, under the conservative assumption that the duration of biochar's effect is only one year. If biochar's effect on soil emissions is longer than one year or even lasts for its entire lifetime, the role of nitrous oxide avoidance would become more substantial and potentially a major driver for the total life-cycle emission reductions of biochar systems.

Reduction in Soil Emissions

In addition to reducing the total need for nitrogen additions to soil, biochar may also affect the soil processes that lead to the production of nitrous oxide. For instance, a study by Taghizadeh-Toosi et al. (2011) found that biochar application allowed for a 70 percent reduction in nitrous oxide emissions from cattle urine patches over the course of the 86-day study period (further research is needed to understand the persistence of the emission reductions). Reductions in nitrous oxide emissions from soils after biochar additions observed by van Zwieten, Kimber, Morris, Downie et al. (2010) were suggested to be due to other changes in the soil that resulted in the release of nitrogen gas (N_2) instead of the GHG nitrous oxide, such as improvement in aeration or a pH shift. This may explain why large reductions in nitrous oxide emissions were reported in some instances (Kammann et al. 2011; Taghizadeh-Toosi et al. 2011), while other studies showed no effects or emissions that varied with time (Singh et al. 2010) or soil water content (Yanai, Toyota, and Okazaki 2007). Research on methane emissions from biochar-amended soils is progressing slower, but shows similar trends to nitrous oxide emissions (Haefele et al. 2011; Karhu et al. 2011).

Increased Nonbiochar Soil Carbon

Another important indirect GHG-related effect of biochar is related to plant response: the change in above- and below-ground biomass (and hence stored carbon) that results from the application of biochar. As discussed above, the potential for enhanced biomass growth only exists when biochar is able to address a plant growth constraint. Therefore, improved plant response will likely be more easily achieved in degraded soils, and soils with high sand content, low pH, and a large proportion of highly weathered minerals (iron and aluminum oxides and kaolinite). Also, biochar additions have been associated with significant reductions in disease occurrence in crops (Elad et al. 2010; Elmer and Pignatello 2011) and yield increases through stimulation of growth-promoting microorganisms (Graber et al. 2010) or plant hormones (Spokas, Baker, and Reicosky 2010).

Both plants and a range of biological processes in soil respond to the introduction of biochar, in turn increasing nonbiochar soil carbon stocks. Soil carbon stocks are highly significant: Batjes (1996) estimates that the first meter of depth of the world's soils contains about 1,500 gigatonnes of carbon, which corresponds to about twice the amount of carbon present in the atmosphere (IPCC 2007). However, the interaction between biochar, nonbiochar soil organic carbon, and decomposition is not well understood. As discussed previously, results from various studies suggest that adding biochar to soil reduces the decomposition of nonbiochar soil organic matter in the long term (for example, Bruun, El-Zahery, and Jensen 2009; Kimetu and Lehmann 2010; Kuzyakov et al. 2009;

Spokas et al. 2009; Zimmerman, Gao, and Ahn 2011). Some studies present interesting explanations for this effect. Novak et al. (2010) combined biochar with powdered switchgrass, which the authors conclude may have shifted the preference of soil microbes from "resident" soil carbon to newly added switchgrass carbon. However, others observe a short-lived "priming effect," where pH, labile carbon, and nutrient additions cause a short-term increase in soil organic matter decomposition (Hamer et al. 2004; Wardle, Nilsson, and Zackrisson 2008; Zimmerman, Gao, and Ahn 2011).

Potential Sources of Leakage and Risks to Emission Reduction

From a strictly climate change mitigation perspective, the risks related to biochar lie primarily in the negative feedbacks that may occur directly or indirectly during biochar production and application. Emissions of methane and soot may occur during inefficient pyrolysis, and native soil organic matter may be degraded after biochar application on unsuitable soils. The risk of indirect land use change from increased pressures on biomass is also important and challenging to quantify. Issues of permanence also arise, particularly related to site erosion and fires.

In theory, all of these risks may be avoided with appropriate standards and safeguards throughout the biochar production chain. Such standards are currently lacking, but their progressive definition and possible implementation—in particular through carbon market protocols and certification schemes—could well rule out certain biochar systems. For example, the slash-and-char model (an alternative to slash-and-burn agriculture) raises some concern in terms of methane emissions during pyrolysis in inefficient kilns. With traditional charcoal kilns used in Kenya and Brazil (Pennise et al. 2001), for instance, emissions of methane from slash-and-char are expected to surpass emissions from biomass burning by 1–6 grams per kilo of dry feedstock. The net GHG impact of the slash-and-char strategy may still be positive, but this risk has to be taken into consideration.

Source of the Biomass Feedstock

One important feedback effect is the risk of indirect land use change with biochar production and application. Biochar is too small of an industry today to be causing substantial indirect land use change. However, it could become an issue in the future if biochar increases overall demand for biomass or changes the relative economic value of certain crops that are particularly suited for biochar production. The risk of indirect land use change can be minimized as long as biochar producers use only "true wastes"—feedstocks that, if pyrolyzed, would not cancel out any positive ecological process (compared to the baseline scenario). Living biomass should only be used as feedstock if it is truly renewable, with no pressures on living biomass elsewhere. As Searchinger (2010) describes:

> Biofuels [or biochar] can only reduce greenhouse gases if the biomass results from "additional" carbon capture. Additional carbon means carbon that would otherwise be in the atmosphere if not incorporated in biomass used for fuel. The carbon must be captured either through additional plant growth or by saving biomass from being broken down through some other pathway.

Under certain circumstances, the changes in land use induced by biochar can also be positive from a climate and ecological perspective. Biochar could be widely beneficial for ecosystem preservation if the gains in agricultural productivity where biochar is applied reduce the need to expand cultivated areas. Again, this is an issue for which the presence or absence of agronomic benefits is crucial to the climate impact of biochar.

Soot and Aerosol Production

Although not taken into account in the Kyoto Protocol to the United Nations Framework Convention on Climate Change, soot is believed to be only second after carbon dioxide as a global warming factor (see, for example, the discussion in Molina et al. 2009). Unlike carbon dioxide, soot is very short lived in the atmosphere, but its warming effect can last much longer when it deposits on ice, which can lead to a dramatic change in albedo, causing rapid melting of ice surfaces in the cryosphere—the snow-capped mountain ranges, brilliant glaciers, and vast permafrost regions. Biochar does not include soot, but because some biochars can easily fractionate into small dust particles, the question must be asked: Can char dust created during applications of biochar be lifted into the upper atmosphere, where it can be transported to fall out onto the cryosphere's glaciers and icecaps?

Aerosols of concern from biomass burning are typically of 2.5 micrometers or less (Brock et al. 2011). Particles larger than this can become airborne for short periods of time, but they tend to be deposited very close to the source. However, larger particles may be transported long distances under exceptional circumstances, for example during dust storms, when particles can be lifted to high altitudes in conditions of extreme wind and dry soil. The critical question, then, is related to the size distribution of biochar particles. This would vary depending on the feedstock and production conditions used, but biochar dust particles are likely to be much larger than the combustion particles (that is, soot).[4] Thus, biochar particles, as compared to the smaller soot particles generated as a product of incomplete combustion, are not expected to be routinely transported beyond the local environment. Appropriate soil conservation tillage methods help to reduce risk of wind erosion during and after biochar application (Blackwell, Riethmuller, and Collins 2009). Such soil conservation practices should be followed wherever conditions favoring wind erosion exist, whether or not biochar is applied to soil.

Indeed, if not properly incorporated into the soil, a fraction of the biochar could possibly become airborne and potentially turn into another climate problem (see discussion above), but at minimum would reduce efficacy as a soil amendment and pose concerns over the health and safety of farm operators. However, a number of simple alternatives to tillage exist, such as humidification or mixing of biochar with manure before surface application or concentrated biochar application in planting holes. In addition, because biochar is so stable, systems could be envisaged where it is not applied every year, limiting the number of times the soil would need to be tilled to incorporate it.

Methane and Nitrous Oxide Production

A potentially more significant and immediate effect is avoided methane and nitrous oxide emissions in situations where biochar would divert wastes that would normally decompose anaerobically, releasing methane and nitrous oxide during this process. If pyrolysis can avoid methane and nitrous oxide emissions when the baseline is anaerobic decomposition, as in landfills, it can also *increase* such emissions compared to full combustion. Thus, the baseline scenario is critical. For example, methane emissions from inefficient charcoal production are a well-known concern, and the Clean Development Mechanism of the Kyoto Protocol contains a specific methodology to curtail them (UNFCCC 2006). In the case of biochar production, this risk does not feature prominently in the literature, as it is assumed that the methane produced during pyrolysis will be combusted and used as a source of renewable energy (including the energy required to run the pyrolysis unit).

However, the question of unintended losses of methane and nitrous oxide during pyrolysis has to be quantified more rigorously and poses a significant research gap. Virtually no scientific studies have been published on this topic, and projects are only now being initiated to fill that knowledge gap.

Permanence

Thus, while microbial and abiotic decomposition of biochar to carbon dioxide will occur, the longevity of the carbon storage potential of biochar would certainly suffice to address permanence-related considerations, as currently required by afforestation and reforestation projects generating carbon credits.

There are studies that predict a half-life of less than a few hundred years (Spokas 2010), though some looked at chars formed under uncontrolled wildfire conditions that likely contained considerable amounts of instable material (Woolf et al. 2010). Moreover, short-term biochar loss in the field may not necessarily occur solely through decomposition, as a significant fraction may have simply been displaced by erosion or transported deeper into the ground by water leaching and soil fauna (Bird et al. 1999; Lehmann et al. 2008; Nguyen et al. 2008). Major et al. (2010), for instance, observed that 45 percent of applied biochar was eroded and only 2 percent decomposed after two years. However, other studies attest to the long-term stability of biochar, finding biochar remnants in soils that are hundreds to thousands of years old to show a chemical composition identical to fresh biochars (Lehmann et al. 2008; Liang et al. 2008). Studies of ancient charcoal deposits face tremendous methodological difficulties in distinguishing between char decomposition and disappearance from erosion, while studies of young chars have a limited time horizon and must resort to unreliable extrapolations from initial rates of decay and artificial ageing (Lehmann et al. 2009).

Although biochar seems to have the potential for stable carbon storage compared to most other land-based climate change mitigation options, risks related to permanence deserve attention. In most cases, biochars can be produced in a way that microorganisms and abiotic processes will take centuries or millennia to decompose half of the carbon compounds of biochar incorporated in the soil.

However, more research is needed to understand how these processes may accelerate under some specific circumstances. It is not well quantified whether oxidation of biochar carbon to carbon dioxide upon entry to river systems can pose a real risk to permanence, or whether decomposition decreases as indicated by enrichment of biochar-type compounds along transportation pathways from rivers to oceans (Golding, Smernik, and Birch 2004; Mitra et al. 2002). Erosion and deposition within the landscape appears to decelerate carbon decomposition in general (Van Oost et al. 2007), but would need to be verified for biochar. In light of current uncertainties and concerned about the possibility of field monitoring of biochar amendments, De Gryze, Cullen, and Durschinger (2010) proposed to limit biochar applications to nonsloping lands in order to reduce risks of lateral erosion and transport of biochar away from its point of application. Erosion is indeed posing a great challenge for measurement, reporting, and verification of the climate impact of biochar projects, which is explored in more detail in chapter 6.

Loss of biochar and of its carbon may also occur prematurely through fire, whether intentional or unintentional. Fire during storage and transportation is a common hazard and has to be managed using regulations that are already in place (Blackwell, Riethmuller, and Collins 2009). Unintentional loss of biochar through fire is likely to be very limited when biochar has been thoroughly incorporated in or mixed with soil. Verheijen et al. (2010) suggested that large concentrations of biochar in soils may also be prone to the risk of smoldering combustion, but high levels of application should be avoided anyway for other reasons in agriculture. Appropriate application rates for soil improvement of typically less than 1–2 percent by mass (20–40 tonnes per hectare if mixed in the topsoil) are unlikely to combust even in the event of a vegetation fire, but can be a risk if biochar is added on top of the litter layer in forests (Knicker 2007). A more important risk, probably, is related to intentional burning of the char as a source of energy. About half of the energy contained in the original feedstock is still present in the biochar. The line between charcoal and biochar only rests on its use, whether for energy or soil application. (The other difference is that charcoal is exclusively made from wood, whereas biochar can be made from any type of biomass.) This risk of intentional biochar combustion disappears once it is incorporated in soil, but appropriate incentives must be in place before this decision is made.

Life-Cycle Analysis of Climate Impacts
In addition to the carbon stored in the biochar itself, a number of important processes can enhance or reduce its climate change mitigation potential. These feedback effects need to be accounted for in order to assess the actual net benefits that specific biochar systems can deliver to mitigate climate change. In other words, there is a move from the narrow perspective of climate change mitigation through direct carbon storage, through biochar production and addition to soils, to a broader view of all the direct and indirect effects of biochar on the GHG balance. Assessing all the impact flows of a product or system is here done in a narrative manner, while chapter 5 will attempt to quantify (most of) these processes through life-cycle assessment methods.

Role of Life-Cycle Assessment

The overall net climate impact of biochar can only be assessed through full life-cycle assessment (LCA) that takes into consideration the indirect effects listed above, as well as other secondary sources of GHG emissions (for example, transport, energy needed to start the pyrolysis). Some biochar GHG assessments and LCAs have been published (Gaunt and Cowie 2009; Gaunt and Lehmann 2008; Hammond et al. 2011; Roberts et al. 2010), but due to methodological constraints and lack of data, none is comprehensive enough to include all aspects discussed above (for example, changes in albedo). They nonetheless tend to take into account, at least to a limited extend, all processes expected to be most important from a climate mitigation point of view (stable carbon in soil, indirect land use change, renewable energy generation, avoided methane and nitrous oxide emissions from landfill, reduion of black carbon (BC) settlement in the cryosphere, emissions from transport, construction of the pyrolysis equipment). However, two dimensions that are potentially quite significant tend to be not sufficiently covered: possible combustion emission as part of the pyrolysis (for example, soot) and potential gains in plant growth and agricultural productivity. Another feature of this first generation of biochar LCAs is that they all took biochar systems specific to temperate developed countries with large-scale pyrolysis units (Australia, the United Kingdom, and the United States). The LCAs presented in chapter 5 of this report intend to respond to the lack of LCAs on biochar systems specific to (tropical) developing countries, with a focus on smaller-scale pyrolysis units.

With these caveats in mind, it is worth noting that the few LCAs published all indicate that the mitigation potential of biochar is greater than the mere recalcitrant carbon stored in the soil, except when negative land use change is involved. The lower-range estimates of most studies converge quite well toward a saving of 1 tonne of carbon dioxide equivalent (CO_2e) per tonne of dry feedstock converted into biochar (with no indirect land use change).[5] A significantly higher degree of mitigation is possible when avoided methane emissions from biomass decay are accounted for (Gaunt and Cowie 2009). Roberts et al. (2010) explain that of this total GHG abatement potential, about 60 percent is attributable to the stable carbon sequestered in the char (which is in line with the rule of thumb of 0.2 tonne of biochar carbon per tonne of dry feedstock explained in figure 3.2), and 30 percent to avoided fossil fuel consumption (using the renewable energy released through pyrolysis). Reduced nitrous oxide emissions from the soil after biochar amendment is however likely to be limited if the soil effect is short lived (2–4 percent of avoided GHG emissions, according to the same source).

When there is indirect land use change associated with biochar production, however, the picture may look completely different. Roberts et al. (2010) calculated that if biochar is made from purpose-grown switchgrass on U.S. cropland (expected to displace agricultural production elsewhere), the overall balance could be as high as 0.353 tonnes of CO_2e emitted—not sequestered—per tonne of feedstock used for biochar.

Figure 3.3 Impact of Biochar on Climate Change Mitigation

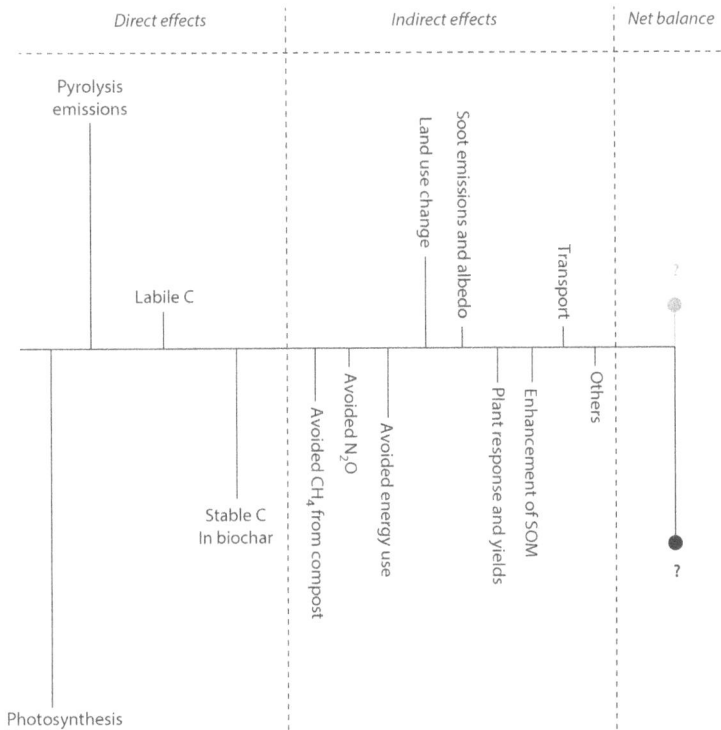

Source: World Bank.
Note: C = carbon; CH_4 = methane; N_2O = nitrous oxide; SOM = soil organic compound matter.

As mentioned earlier, the net emission reductions from a biochar system are, under most assumptions, only superior to a full combustion of the biomass if there are soil benefits that yield additional emission reductions. These could arise from improved growth, reduced fertilizer requirements, or lower methane or nitrous oxide emissions from avoided landfill or from soils that received biochar. If no soil benefits are achieved, the emission reductions of a biochar system are likely not greater than those of a bioenergy system using combustion (Roberts et al. 2010; Woolf et al. 2010).

Figure 3.3 summarizes the overall impact of biochar on climate change mitigation (GHG emissions and other processes such as changes in albedo) as a combination of direct and indirect effects. This overall impact can be positive or negative, depending on the characteristics of specific biochar systems.

Time Dimension

The net climate impact of biochar also varies with time. LCAs are static snapshots of a situation at a certain point in time, and unlike system dynamics modeling, which uses systems of equations representing stocks and flows modeled dynamically over time, temporal dynamics are not included. However, a timeframe can be included that is specific to the product or system analyzed and that

Figure 3.4 Alternative Scenarios for Biomass Carbon Dynamics

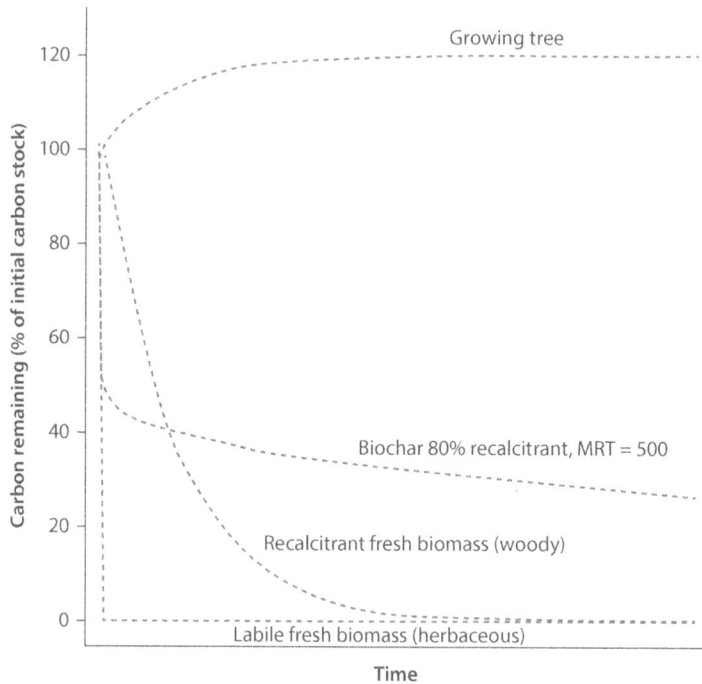

Source: Whitman, Scholz, and Lehmann 2010.
Note: Each curve represents the fate of an equivalent mass of organic matter. MRT = mean residence time.

represents the impacts at the end of that period. The selection of this timeframe will influence the net impact of the system. If a series of these calculations is done with different timeframes, the development of the emissions balance can be studied over time (still, no feedbacks can be included in such a serial LCA). In the first months, or years, after pyrolysis, biochar does not compare well with alternative scenarios in carbon terms. Indeed, pyrolysis releases instantaneously about 50 percent of the carbon contained in the original biomass, which would have taken longer to decay naturally (figure 3.4). If the baseline scenario is biomass burning or fast decay of labile biomass, biochar production quickly results in increased carbon storage, while for slowly decaying organic matter, biochar can pay back its carbon debt only after a period of time. It is generally agreed that land-based mitigation strategies with payback periods of more than 10 years (which is the case for various biofuel models) should be avoided (Woolf et al. 2010).

In the longer term, another effect comes into play: the positive feedback that a biochar cycle can cause on the next, through increased plant growth and availability of biomass residues for future pyrolysis. (This aspect has been well understood by proponents of the slash-and-char strategy.) The contribution of such a virtuous cycle to the overall mitigation potential of biochar is hard to quantify—and is once again very context specific—but should not be neglected. The system dynamics model developed by Whitman et al. (2011) to predict the effect of the introduction of a biochar-producing cookstove into a western Kenyan household on the GHG impact of the system demonstrates one way to quantify this effect. In the model, based on field data, biochar application to soils increases maize yields, which, in turn, results in more crop residues available to be used as stove fuel, replacing unsustainably harvested fuelwood. However, this effect occurs gradually as the stove produces biochar over the years. The opposite effect may occur if biochar accrual in soil reaches levels that do not further improve productivity or even reduce it; similar arguments can be made for nitrous oxide emissions from soil. At some point in time, more biochar accumulated in soil will not result in a linear increase in emission reductions. Without a time dimension to modeling efforts, it is very challenging to quantify such nonlinear effects.

Potential Global Climate Change Impacts of Biochar at Scale

The previous discussion assessed the net impact of turning 1 tonne of biomass into biochar and using it as a soil amendment. If biochar were brought to scale, what could be the size of its contribution to global mitigation targets? This discussion does not allow determination of the scale at which biochar systems *should* be implemented, but rather provides guidance as to whether biochar has the potential to be a real contributor to climate change mitigation on a global scale. If biochar has the potential to be important globally, each project must still be assessed for sustainability and net climate impact on an individual basis, and how any wider-scale implementation would impact the global and regional social, biological, and economic systems discussed elsewhere in this report.

Woolf et al. (2010) examined the theoretical (or "technical") global mitigation potential of biochar within the boundaries of sustainability criteria. To avoid endangering food security and ensure habitat and soil conservation, the authors restricted their analysis to clearly identified waste feedstocks (for instance, paddy rice straw not used for animal feed) or biomass generated through limited land use changes that do not reduce agricultural production (for instance, transition from pasture to silvopasture). The availability of sustainable feedstocks is clearly identified as the limiting factor for a large-scale deployment of biochar, not the availability of suitable land (at least for the foreseeable future).

According to their calculations, annual net emissions of carbon dioxide, methane, and nitrous oxide can be reduced by a maximum of 1.8 gigatonnes of carbon per year, equivalent to 12 percent of current anthropogenic emissions (half of the avoided emissions are expected from the carbon sequestered in the char, 30 percent from replacement of fossil fuel energy by pyrolysis energy, and 20 percent from avoided emissions of methane and nitrous oxide). It is not possible to estimate at this point what a more realistic potential is, given all existing economic and social constraints, barriers in technology development, and competing uses in a future bioeconomy. The cited potential merely serves here as a justification for further inquiry about biochar's utility on a project level.

Many biochar proponents emphasize that biochar deserves particular attention in a mix of strategies (or wedges; Pacala and Socolow 2004), as being one of only a few options that can actually withdraw carbon dioxide from the atmosphere. The overwhelming majority of mitigation technologies, including carbon capture and storage in the coal industry, propose to limit present and future emissions only, but cannot reduce the excess carbon dioxide in the atmosphere. As a threshold of cumulative emissions is likely to be reached beyond which no level of emissions can be considered safe, strategies that can actually absorb carbon dioxide from the atmosphere will become more essential than others (Woolf et al. 2010). The only really available "technology" that at present captures carbon dioxide from the atmosphere on a large scale is photosynthesis. Biochar can slow down the decay of this captured carbon dioxide in plants that would otherwise be fully returned to the atmosphere. Contrary to other land-based mitigation strategies, such as reforestation, no tillage, or reducing emissions from deforestation and forest degradation (REDD), which also have the advantage of drawing down atmospheric carbon dioxide, biochar does not face so strongly the issue of permanence; it is harder to reverse the effects of biochar systems compared to forestry systems, which are particularly prone to the risk of destruction through fire or diseases. In comparison to full combustion of biomass, however, biochar systems do face greater variability in a portion of their emission reductions, because they accrue or have to persist over time (such as greater plant growth or lower GHG emissions from soil). On very long timescales of several hundred to thousands of years, the disadvantage of biochar lies in the very fact that biochar will eventually decompose, albeit more slowly than the biomass that it was produced from (see above).

Social Impacts

The Roundtable on Sustainable Biofuels fifth principle states, "In regions of poverty, biofuel operations shall contribute to the social and economic development of local, rural and indigenous people and communities" (RSB 2010). This guiding principle is also essential to the success of any biochar project. Because biochar systems can affect energy, health, economics, and food security, among other major issues, it is absolutely necessary that care be taken in their design, particularly in developing countries.

Health and Labor Impacts

The institution of biochar systems has potential positive impacts on the health of system users, particularly in the case of biochar-producing stove projects. However, there are also risks associated with such systems. These various health impacts will be investigated in the following subsections. They could occur throughout the production process, and include decreased fuel-gathering pressures, reductions in indoor air pollution, risks associated with handling of biochar and inappropriate conversion technology and operation, and the effects of improved crop yields.

Increased Fuel Efficiency and Decreased Fuel-Gathering Burden
Along with potentially decreasing indoor air pollution, decreased fuel demand has been an important driver for the development of improved cookstove projects. While wood gathering for fuel may not be the primary reason for

deforestation in many at-risk areas (Defries et al. 2010), it may still be an important factor in some regions of Africa (Fisher 2010), and it is certain that the challenge of gathering sufficient wood for cooking in a wood-constrained region is substantial. It is usually the women in developing countries who spend 4–14 hours per week collecting sufficient fuel for their household needs (WHO 2000). Decreasing the burden on women through cookstove improvements could be substantial. In the case of biochar-producing cookstoves, these improvements could come both from increased fuel efficiency, similar to improved combustion stoves (Johnson et al. 2008; MacCarty et al. 2008), and more importantly from expansion of potential fuel sources (Torres et al. 2011). The wider variety of feedstocks for pyrolysis and gasification stoves greatly enhances opportunities to satisfy on-farm fuel needs and reduces labor required to collect wood (Torres et al. 2011; Whitman et al. 2011). Together with the biochar production for soil improvement, the reduction in wood gathering may be an important driver for adoption. Equally important is the flexibility of biochar-producing stoves, not only in giving wider access to different types of fuel, but also in allowing the user to choose how to use the biochar. Depending on the stove design, biochar can be either further combusted in the stove for additional cooking energy or removed from the stove and saved. Once removed, the user can then have the options of returning the biochar to the stove to use as fuel, applying it to soil, or using the biochar for applications such as water filtration or sanitation.

Research on energy efficiency and emissions of biochar-producing stoves is still in its infancy (Roth 2011; Torres et al. 2011). It is important to recognize that if decreasing absolute fuel use in addition to wood use is the ultimate goal, improved combustion cookstoves would be a better choice than biochar-producing cookstoves. Because as much of the fuel biomass as possible is combusted for energy in improved combustion cookstoves, and biochar-producing cookstoves explicitly aim to retain some of the biomass in the form of biochar, it would be technically impossible for a biochar stove to gain as much energy from a given amount of biomass as an improved combustion cookstove could (Johnson et al. 2008). However, biochar cookstoves may be substantially more efficient than traditional cookstoves, and if the soil or climate benefits of biochar are valued, then the biochar-producing stove may be preferred over a combustion stove.

Reduction of Indoor Air Pollution
Every year, indoor air pollution causes nearly 2 million deaths (WHO 2011). Since improved cookstove projects began, one of the major impetuses for developing and promoting improved cookstoves in developing countries has been their potential to reduce indoor air pollution (Hyman 1985; Raju 1954). While complete combustion of hydrocarbons will produce only carbon dioxide and water, when sufficient oxygen is not available, harmful products of incomplete combustion, including carbon monoxide and small respirable particulates, are produced. These compounds are associated with harmful effects, including pneumonia and other acute lower respiratory infections, chronic obstructive pulmonary disease such as chronic bronchitis, and systemic effects as carbon monoxide

limits the oxygen-carrying capacity of the blood (WHO 2011). The effects of products of incomplete combustion are exacerbated when cooking is carried out indoors, without adequate ventilation. These negative impacts affect women and children disproportionately. Women in many societies are responsible for the cooking, and spend between three and seven hours each day near the stove preparing food, often with their children close by. Because of this, most deaths attributable to indoor air pollution occur in females (WHO 2011).

The development and introduction of cleaner-burning and more efficient cookstoves has been promoted for decades to address this problem, with varying degrees of success (Barnes et al. 1993). Stoves have been shown to reduce the production of products of incomplete combustion both on a per unit of biomass burned basis and on a per unit cooking energy basis (Johnson et al. 2008; MacCarty et al. 2008; Roden et al. 2009). For example, MacCarty et al. (2008) found that the relative emission of products of incomplete combustion for an improved gasification stove (which would have similarities to a pyrolysis cookstove) in laboratory testing was about half that of the traditional three-stone cookstove. While air pollution concerns are greatest surrounding cookstoves, it should also be noted that modern gasifiers and improved charcoal kilns that flare or recycle off-gases produce fewer emissions than traditional charcoal kilns (Brown 2009). While it is reasonable to expect that improved biochar systems have the potential to be cleaner than traditional methods, data comparing them to other improved cookstoves or kilns are currently lacking. Extensive emissions testing of biochar cookstoves is currently under way, and the continued development of these stoves will naturally focus on reducing indoor air pollution in addition to reducing GHGs. For example, a recent World Bank report (*On Thin Ice: How Cutting Pollution Can Slow Warming and Save Lives*, 2013) shows that improved biomass (wood) and coal heating stoves could save about 230,000 fewer lives annually with the vast majority of these health benefits occurring in Organisation for Economic Co-operation and Development (OECD) countries.

Of course, the positive effects of improved cookstoves will only be realized if the new technologies are adopted. Key elements suggested for successful stove adoption in various improved cookstove programs include educating women on the harmful effects of smoke and emissions, basing stove designs on traditional forms, targeting specific stove types to areas with different needs (for example, portable or not), and creating business opportunities for local stove makers and intermediaries (Aggarwal and Chandel 2004; Barnes et al. 1993; Hyman 1985).

Potential Negative Health Impacts Due to Particulate Black Carbon and Toxins
In the process of producing, storing, transporting, and applying biochar, there are several potential risks to human health. The primary areas of concern are black carbon emissions or charcoal dust, silicon dust, and toxins—particularly polycyclic aromatic hydrocarbons, dioxins, and furans. No peer-reviewed data that investigate the risks of toxins and black carbon or charcoal dust for biochar systems are available, but some reports provide insights.

As described above, one of the major potential benefits of improved cook-stoves is the reduction of indoor air pollution from compounds such as carbon monoxide and respirable suspended particulate matter (Kanagawa and Nakata 2007). While this benefit could be realized during the cooking process, care would have to be taken to ensure that risks from handling the fine particulate matter of the biochar product are minimized. Small particles (less than 10 micrometers in diameter) are thought to be the most dangerous for human health, and studies often focus on particulate matter less than 2.5 micrometers in diameter ($PM_{2.5}$), which is capable of penetrating deep into the lung, causing severe respiratory effects (WHO 2000).

Similarly, charcoal dust is another potential area of concern where biochar is being produced in traditional ways or handled by workers. Workers who produce charcoal, process it into briquettes, or bag it for distribution are routinely exposed to charcoal dust, yet there is very little published literature on the impact of exposure to charcoal dust. The primary identified occupational hazard among workers in traditional charcoal industries is smoke inhalation (Kato et al. 2005). However, De Capitani et al. (2007) examined three cases of pneumoconiosis (black lung) in wood charcoal workers and found that highly exposed workers accumulated carbon dust in the lungs over long periods. The authors conclude that, "Despite the few cases published so far, pneumoconiosis due to wood charcoal might be an underestimated occupational risk, and early diagnosis and prevention must be addressed mostly in developing countries, where low industrial hygiene standards might expose workers to dust above threshold limits." In systems where biochar is being produced using traditional kiln technology, this risk should be monitored and appropriate health and safety precautions should be taken. Strategies would also need to be developed to ensure that practitioners are not exposed to particulate matter risks during biochar handling or application to soils. These strategies could include keeping it covered during storage and transport and wetting the biochar during its application to reduce dust formation. Methods for containing dust from charcoal that is being crushed for briquetting have been developed (Thomas 2008), and could be used by farmers needing to crush biochar prior to field application.

The knowledge of health impacts from dust that contains silicon fibers is much stronger—silicosis is a serious health impact for workers in many industries that process materials containing silica, such as ceramics. Even dust from crop wastes high in silica, such as rice husks, can cause a silicosis-like syndrome in rice mill workers (Lim et al. 1984). Rice husk biochar also contains silica, and during pyrolysis or gasification silica could crystallize, increasing its threat to health. The transition of silica in rice husk from the amorphous form to the crystalline form generally takes place at temperatures above 850°C, and heating time is also a factor. More investigation is required, but pyrolysis and gasification below this temperature threshold may not be likely to produce much crystalline silica (Shinohara and Kohyama 2004).

Polycyclic aromatic hydrocarbons (PAHs) are chemical compounds that are produced as by-products of fuel burning (whether fossil fuel or biomass). Some

PAHs are carcinogenic to humans, but many are not. PAHs are found naturally in soils as a result of wildfire, and many microbes are able to metabolize them. PAHs have been identified in some biochars (Brown et al. 2006; Jones, Lopez Capel, and Manning 2008), but the conditions under which they form and how pyrolysis systems could be designed to prevent them remain to be clarified. Karve et al. (2011) analyzed PAH levels in several biochars, including carbonized rice husk produced by a number of rice husk gasifiers operating in Cambodia, and found one species of PAH that exceeded levels of concern. However, this may not be critical, given that (a) the biochar would be diluted in much greater volumes of soil, and (b) the PAHs in biochar may not be leachable or plant available due to being strongly sorbed by the biochar.

Dioxins are predominantly formed at temperatures above 1,000°C. Most pyrolysis technology operates well below that temperature. However, any proposed high-temperature pyrolysis or gasification technology should be assessed and monitored for possible dioxin production. A low-temperature pathway for dioxin also exists (the de novo pathway), which requires the presence of oxygen and chlorine. Because pyrolysis operates under no or low oxygen contents, this pathway is also unlikely in biochar systems, but feedstocks that contain significant amounts of chlorine or metals are more prone to dioxin production (Garcia-Perez 2008) and need to be scrutinized.

Increased Labor under Slash-and-Char Systems

The practice of converting swidden, "slash-and-burn" agricultural systems to "slash-and-char" systems has been suggested as a means of improving soil fertility and reducing carbon loss from the system (Lehmann et al. 2002). The case study described by Lehmann and Waddington (Lehmann and Joseph 2009) involves wood being hauled from the cleared area to be charred in buried or improved kilns before the remaining small biomass is burned. The biochar could then be sold as charcoal fuel, or returned to the soil, where even a small, 2–3 percent increase in yield of high-value pineapple and annatto could offset biochar production costs in one year (Lehmann and Joseph 2009). However, this process could require substantially more labor, with an estimated 10.5 person-days per hectare for gathering the woody material, 6.8 person-days for building the kilns, and 2.8 person-days to return the woody material to the soil. This increased labor, for people who already work very hard, may not be an acceptable change unless the payback is substantial.

Nutrition Effects from Improved Crop Yields

Soil degradation is strongly linked to human nutrition and health (Lal 2009) and farmers and pastoralists of Asia and Africa make up nearly two thirds of the world's hungry (Borlaug 2007). If biochar improves crop yields or crop resilience (as discussed in section "Impacts on Soil Health and Agricultural Productivity" above), the institution of biochar systems could help buffer practitioners against hunger. Whether or not the application of biochar to soils results in an alteration of the nutrient content of crops, an increase or maintenance of yields under environmental stresses such as climate change could be critical for smallholder farmers, for example those in Sub-Saharan Africa, where yields of staple crops are projected to decline by between 5 percent and 22 percent by 2050 as a result of climate change (Ringler 2010). As discussed above, targeting biochar to soils and agricultural management systems may—if appropriately applied—help ensure that these benefits are realized.

Access to Energy through Biomass

Energy Implications of Biochar

Substantial portions of the world currently rely on inadequate energy supplies and energy forms for even basic household needs such as lighting, heating, and cooking. In 2001, the World Bank reported that 1.6 billion people used no modern fuels (coal, kerosene, electricity, natural gas, or liquefied petroleum gas), and that populations in developing countries consumed 5 percent of the modern energy that those in developing countries consumed per capita (Energy and Mining Sector Board 2001). The introduction of a biochar system to a region would be expected to create changes in the accessibility of biomass and energy. This is critical, because energy consumption is strongly correlated with national income as well as human development (Energy and Mining Sector Board 2001). For small-scale biochar cookstove projects, this improved access to energy could yield the many positive improvements related to health impacts described above, while for larger-scale biochar energy projects, the provision of energy could supply lighting after sunset (extending the workday or time during which children can concentrate on schoolwork), power machines to increase productivity, and provide energy for critical functions such as the refrigeration of vaccines or water pumping. More broadly, the ability to generate income through nonagricultural activities such as microbusinesses is limited by energy, among other factors (Kaygusuz 2011).

   While all these energy services are valuable, for a given project it would be important to consider which kinds of services should be targeted and which groups would benefit from those services. The prioritization of energy services would be expected to differ from region to region, and should be determined in consultation with the community or groups involved.

Barriers to Success in Biochar Energy Projects

A case study of the implementation of a gasifier system fueled by fast-growing tree biomass (with parallels to a midscale pyrolysis unit) in Vanuatu found that it ran successfully for several years, providing electricity to a school (Woods, Hemstock, and Burnyeat 2006). Its ultimate failure was due to a lack of technical support and barriers imposed by external agencies rather than local social or environmental factors, suggesting that the development of a long-term strategy for supporting new energy systems may be critical for ensuring their success. In the same report, the authors noted that highlighting the fact that local energy needs can be met locally would help promote "ownership" of biomass energy strategies on a local to regional basis.

While the issues surrounding alternative uses of biomass are investigated in section "Competing Uses of Biomass" below, these dynamics would be expected to influence certain groups of people in different ways, with the potential for systematic marginalization of vulnerable groups. Furthermore, when biomass is being used to produce bioenergy for essentially an unlimited demand (as opposed to demand for energy for a relatively fixed use, such as daily cooking needs), the likelihood of unsustainable biomass use or negative social impacts is increased. A report from the International Institute for Environment and Development (Nhantumbo and Salomão 2010) examined biofuel projects in Mozambique as a case study for understanding the impacts of biofuels on the livelihoods of the rural poor. The study found that inadequate planning and failure to implement existing regulatory policies has resulted in conflicts between different resource uses and users. Indeed, some interviewees cited comparatively lax regulatory standards in Mozambique as a reason for the greater rate of biofuel project development there compared to South Africa. This observation highlights the need for international standards and their enforcement. Community consultations have been found to be inadequate to protect the rights of communities, and the promised positive social impacts of projects have been provided on an ad hoc basis, without clear timelines for their implementation. If biochar projects are designed explicitly in the context of community development, rather than economic investments, then this could be a strong step toward avoiding similar issues.

Kaygusuz (2011) stresses the importance of the gender dynamics of energy production and access. Due to differences in access and control over land and productive assets, the risks of biofuels are expected to affect men and women differently (FAO 2008). Women tend to be those responsible for spending long hours or travelling long distances to gather biomass for fuel. Also, women are often responsible for producing food crops in many areas. While altering these activities through a biochar energy strategy could alleviate pressures on women and also provide them with new energy resources, the changing dynamics of biomass or an increase in land clearing for large-scale biofuel production could also disproportionately impact these women (Kaygusuz 2011). That said, household- and community-scale bioenergy projects likely have the potential to improve energy services for basic needs and generate income in rural developing regions while alleviating health and labor burdens, particularly if a participatory approach is used when developing and implementing projects and if gender-differentiated impacts are considered in their design (Kaygusuz 2011).

Benefit-Sharing Issues Arising with the Potential Development of Biochar through Carbon Credit Schemes

The nascent economic topic of potential carbon credits for biochar systems is covered in chapter 6, but there is certainly a social aspect to carbon crediting as well, which is briefly addressed here. Two issues arise: the general principle of using carbon credits generated in regions of the world with the fewest per capita GHG emissions in order to offset the emissions of industrialized countries; and the question of who benefits ultimately financially from the generation of carbon credits.

The only way the first issue can be justified—also from a climate mitigation perspective—is if the activity is "additional" and social and economic development are at the core of the project, so that carbon credits are applied in order to enable access to these positive benefits (Whitman, Scholz, and Lehmann 2010).

In consideration of the entire value chain of a biochar system, where there are a number of different actors throughout the process chain, determining who should receive payments for the carbon credits can become complex. As outlined in table 3.2, reductions in emissions can occur at many points, and their value may be influenced by other parts of the process. For example, the biochar producer is responsible for stabilizing the feedstock biomass as biochar as well as for the initial emissions during pyrolysis. For the producer's efforts to result in net carbon reductions, the farmer responsible for applying biochar to soils must follow the expected procedure (that is, protocol or method) and ensure that the biochar is not burned for energy or applied to inappropriate soils, resulting in net GHG emissions. Thus, it is essential that the project be conceived in its entirety, with the agreement of all participants as to what any carbon financing will be applied to, and ensuring that carbon credits are not double counted. Although the aforementioned complexity of a biochar system might seem challenging, there is—particularly from a social perspective—an inclusiveness factor involved in operating a successful biochar operation particularly at the scale of smallholders. While larger systems might incorporate and combine several process steps for which then only one project entity is responsible, a smallholder setting would suggest that different process steps might be carried out by different actors which will ultimately only be eligible to receive carbon credits if the biochar system in its totality functions as one. As a result, increased cooperation and inclusiveness might be observable along the value chain of a well-operating biochar system. However, it is still too early to fully assess these social effects, given the nascent nature of biochar systems per se.

Competing Uses of Biomass

The availability of biomass is a key part of what defines the potential scope of biochar projects. Precisely which categories of biomass could be most appropriate for a biochar system are highly location and system specific. Many different types of biomass can be used in pyrolysis systems to produce biochar, and different fuel types would be optimal in different systems. Pyrolysis systems that operate at different scales require different amounts and, thus, sources of biomass. Furthermore, even if the biomass is suitable for pyrolysis, it may not be available or affordable. There is often an opportunity cost attached to the use of a specific biomass source for pyrolysis. Biomass management is a particularly sensitive issue for those developing countries with widespread degraded soils and limited biomass resources. There are many factors that limit how much biomass should be considered "available" (table 3.3), all of which must be taken into account for successful biochar projects.

Table 3.3 Potential Biomass Use and Limitations

Potential	Limitations considered
Theoretical	Biological production
Geographic	Existing land area
Technical	Land required for food, housing, infrastructure, and natural areas
Economic	Profitability
Implementation	Social and policy constraints
Sustainable	Food security, habitat conservation, and soil preservation

Sources: Offerman et al. 2011; Woolf et al. 2010.

Here, some common uses of biomass that may compete with biochar production are investigated. Competing uses for a potential feedstock can be predicted to some extent by the broad category the biomass type falls into. This discussion of biomass sources follows the framework applied by Offerman et al. (2011), which distinguishes between energy crops and residues. Energy crops are purpose-grown plants for bioenergy production while residues are divided into three categories: harvest residues, such as straw, leaves, or tree thinning residues; process residues, such as bagasse, manure, or mill wastes; and wastes, such as rotten food products, waste wood, invasive species, or municipal solid waste. None of these categories is exempt from competing uses.

Purpose-Grown Energy Crops

The use of energy crops for biochar production would be expected to face all the same issues as is the case in other bioenergy or biofuel projects with respect to growing dedicated biomass. Briefly, these include the diversion of food crops for fuel, diversion of arable land from food crops (Pimentel et al. 2009), direct and indirect land use change (Lapola et al. 2010), and whether or not energy crops could truly be constrained to degraded or marginal lands (that is, land unsuitable for food crops) (FAO 2008). These issues have been investigated at varying degrees of depth, but all could potentially act as competing uses for biomass feedstocks. Furthermore, while it is feasible that some biomass types could be appropriate for making biochar but not for other standard bioenergy uses, it is likely that there could be competition from other biomass energy systems besides biochar production. The issues surrounding this category of feedstock are highly complex, making it particularly challenging to prevent negative externalities.

Biomass Residues

Composting or Soil Application

Any time soil is deprived of biomass that would have otherwise decayed in situ and hence protected and enriched the soil, there is the potential for loss of nutrients. Dead biomass that remains on soils, whether it is forestry slash, maize residues, or leaf litter, protects the soils from erosion and will eventually decompose, putting at least parts of the carbon and nutrients back into the soils (Lal and Pimentel 2007). If these functions are valued, care must be taken not to excessively remove biomass, and for some easily degradable soils, any biomass removal would put the soil at risk. Using harvest residues for bioenergy production is a clear example of this process—where biomass would have been returned to the soil, it is now being diverted. However, in the case of a biochar system, if the biochar is returned to the soil from which the residues came, many of the mineral nutrients could be returned (with the important exclusion of much of the nitrogen and sulfur—see section "Impacts on Soil Health and Agricultural Productivity" above), and soil carbon stocks could be enhanced (although it is important to differentiate the form and function of biochar carbon in soil from non-BC soil carbon). However, important functions of leaf litter mulches for erosion protection may not be delivered by biochar incorporation to soil. This requires scrutiny and may significantly limit or even exclude the use of crop residues in certain regions at risk from soil erosion.

The use of process residues and waste residues, which may normally be sent to the landfill, may not result in losses to soils, but each scenario would need to be evaluated independently to determine what the standard or baseline practice is. For example, Offerman et al. (2011) designate manure as a process residue, which in a concentrated animal feeding operation may be too abundant locally to apply safely to soils without risk of eutrophication (Bradford et al. 2008), but on a small Ethiopian farm is an important agricultural resource (Duguma, Darnhofer, and Hager 2009). In addition, the choice to apply biomass for one use over another does not have to be permanent, but could vary over time. For example, biochar could be produced from biomass that is normally used as compost once every decade if that would optimize soil nutrient management and fertility. An important question here, as elsewhere, is what the baseline scenario is. For example, if the baseline is application of fresh biomass with a high C:N ratio (such as wood), it may tie up nitrogen during decomposition, making the nutrients unavailable, while turning it into biochar would make the carbon less available to microbes, reducing this effect (Torres 2011).

Animal Fodder
Competition for biomass to be used for animal fodder instead of biofuel systems runs in parallel to those associated with human food–fuel trade-offs, where land or crops previously used for animal feed are diverted for bioenergy production. Particularly important for harvest and some process residues, it is relevant in both developed- and developing-country contexts. For example, in a concentrated animal feeding operation in Iowa, United States, maize plants previously used for silage production for cow feed might be diverted for or made more expensive due to their demand for bioenergy production. In the small Kenyan households studied by Torres et al. (2011), 25 percent of maize residues were reported to be used for animal feed or building materials, leaving 75 percent potentially available for pyrolysis in a biochar-producing cookstove, for which the need for soil protection has to be considered (Whitman et al. 2011). Whereas the Iowan farmer might turn to the grain markets for alternatives, the Kenyan smallholder may be more constrained.

"True Waste"

Despite the fact that biomass is a resource in many scenarios, in some instances it could be considered "true waste" (Whitman, Scholz, and Lehmann 2010). While from a soil nutrient perspective composting municipal green waste would be better than using it for bioenergy, from a climate change standpoint the opposite may be true, and in many situations it is simply landfilled, providing neither climate nor soil nutrition benefits (Lundie and Peters 2005). If the default or business as usual scenario for biomass use would simply be rotting in a landfill without contributing to soil nutrient enhancement or being used as animal fodder and if this true waste could be economically diverted and used for bioenergy production, it could be an attractive option as feedstock for a biochar system. Indeed, in such scenarios the cost of this feedstock might be zero, or even negative in cases where the feedstock provider would have otherwise been paying a tipping fee to dispose of it. However, one challenge associated with using municipal wastes for biochar production is often its high moisture content. Roberts et al. (2010), for example, found that yard waste required the most energy for drying before the pyrolysis process compared to corn stover and switchgrass feedstocks.

Another model of true waste could include biomass that could be used to produce biochar while providing its original function. For example, if traditional swidden agricultural practices could be adapted to use charring instead of outright burning of biomass to clear lands for agriculture (Lehmann and Joseph 2009), then some of the advantages of biochar production and its application to soils could be gained while the original use of the biomass is not compromised. However, the success of such a system would depend on social, economic, and environmental factors specific to each farming system.

The use of true wastes for biochar production could minimize unwanted land use impact and leakage (Whitman, Scholz, and Lehmann 2010). However, the concept of true waste is system defined, as one person's rotting food is another's compost, for example. While the same outcomes of burning—to clear land for hunting or cultivation—may be achieved through a slash-and-char system, other important ecological functions may not be, such as the stimulation of plant growth or germination through fire (Bond and Keeley 2005). In biomass-constrained developing regions, very little biomass can be considered "waste." However, if its alternative use is inefficient burning in a cookstove, then using an improved cookstove that also produces biochar could provide a viable alternative use (Torres et al. 2011). Conversely, as discussed above in section "Impacts on Soil Health and Agricultural Productivity", many biological wastes that are indeed true wastes would also be unsuitable for biochar production and application to soil. Any contamination that is not destroyed through the pyrolysis process, such as heavy metal pollution, would remain in the biochar and could have harmful effects. Thus, it is critical that the value of a true waste does not overshadow the risks of contaminated waste products when selecting a feedstock.

Alternative Bioenergy Production

Whether the biomass in question is an energy crop or any sort of residue, another bioenergy system could compete with biochar production for its use. The critical questions in these scenarios might include what form of energy is being produced and how it serves the community's needs; and if it is serving the same function (as with improved cookstoves replacing three-stone fireplaces), which process is more efficient or best meets other criteria such as indoor air pollution. For example, the analysis conducted by Woolf et al. (2010) to predict maximum global sustainable potential biochar production notes that the extent of their biochar scenarios is incompatible with a simultaneous nonbiochar biomass energy strategy. They propose a mixed approach whereby each system is applied where it would produce the optimal outcome. For example, the effect of biochar on soil fertility and the carbon intensity of the energy it is replacing are key for achieving a beneficial climate outcome, so in regions with fertile soils and where coal combustion would be replaced with bioenergy, nonbiochar strategies may be preferred. Similarly, Whitman et al. (2011) find that an improved combustion cookstove may be preferred over a biochar cookstove in scenarios where fuel use is very high and fuel gathering is more unsustainable.

Risk Assessment

Biomass and even what are considered biomass residues are often important components of elements such as building materials, fuel, livestock feed, direct soil amendments, and soil protection. Regions or locations where biomass sources can be identified where such alternative uses are minimal or unimportant will likely be better suited for biochar interventions. It is expected that biochar will best complement locations characterized by high biomass availability coupled with compromised agricultural yields and soils. In such locations, biochar would be more likely to demonstrate its value quickly enough to be of interest to the people, improving adoption rates. It is essential that any biochar production does not rely on biomass that previously supplied other critical uses. It is important to be aware that these uses may not appear to have high economic value but can in fact be providing essential services.

The concept of true wastes is a useful principle for the development of biochar projects, but should be carefully applied to very specific categories of biomass residues. Key indicators of a true waste biomass may include a net cost for its management, its ultimate fate being combustion without energy capture or decomposition in peri-urban or industrial landfills without contributing to soil nutrients, soil protection, and soil carbon. Its use should not endanger food security, natural habitats, or soil health (Woolf et al. 2010).

Human appropriation of the world's net biomass production is in the vicinity of one quarter of total biomass produced (Erb et al. 2009). The optimal use of a given amount of biomass is always going to be defined by value judgments. Different weighting of the importance of soil health, nutrient cycling, energy production, climate change impacts, economic value, and many other factors will almost inevitably result in different recommendations for its ideal use. Because there is no unbiased "optimal" use for a given biomass supply, transparency in the value judgments that underlie any biomass use decisions is of the essence.

Notes

1. Chan and Xu (2009) reviewed a large number of biochars from a variety of feedstocks and found a mean pH value of 8.2.

2. "Labile" carbon is less resistant to weathering and can readily be broken down and released as carbon dioxide into the atmosphere. "Recalcitrant" carbon—which generally constitutes a much larger proportion of total carbon content of biochar than labile carbon—is much more resistant to degradation.

3. The mean residence time is the average time that biochar carbon remains in the soil before being completely broken down.

4. C.A. Masiello, personal communication.

5. Lower-range estimates (in tCO_2e per tonne feedstock) are 0.86 for Roberts et al. (2010) with late stover in the United States, 1.3 for Gaunt and Cowie (2009), and slightly above 1 for Hammond et al. (2011) with barley straw in the United Kingdom.

CHAPTER 4

Survey and Typology of Biochar Systems

Survey

The use of biochar is a relatively new technique in modern agriculture, building on a body of research that is mostly less than a decade old. Dedicated institutional capacity to research and develop biochar applications is only now beginning to emerge. A survey was undertaken to obtain an overview of the status of biochar projects globally, particularly in developing countries. The survey had four purposes: (a) to provide a snapshot of the types of biochar projects that currently exist in developing countries; (b) to gather information about constraints and opportunities in these biochar systems; (c) to develop a typology based on the survey results; and (d) to select a few projects to study the life-cycle greenhouse gas (GHG) emission reductions. The survey was conducted by using the existing electronic network of the International Biochar Initiative (IBI). With 3,500 newsletter subscribers from 113 countries, IBI has become the nexus of a fast-growing multidisciplinary network of people interested in biochar, five years after its creation at a side meeting held at the 18th World Congress of Soil Science, Philadelphia, United States, 2006.

In order to cast the widest possible net, the request was also posted at various online discussion forums for biochar and for cookstoves (recently, biochar-making cookstoves have become a topic of interest to cookstove developers). In response, IBI received 154 completed surveys from 41 countries (figure 4.1).

The majority of the projects identified themselves as still in the beginning stage. Only 39 projects were self-identified as mature or having measurable results. Of these, 12 projects were identified as potentially having enough results and sufficient data quality for a life-cycle assessment (LCA). These projects were contacted in January 2011 and were sent a follow-up questionnaire to collect additional project data.

An additional survey was drafted and invitations were sent to those who responded to the initial survey in order to learn more about various social and cultural barriers to biochar adoption. This survey had 48 responses to questions about barriers, existing indigenous biochar use, how projects cope with limited supplies of biochar, and project reliance on carbon financing. The results of this

Figure 4.1 Distribution of Project Locations

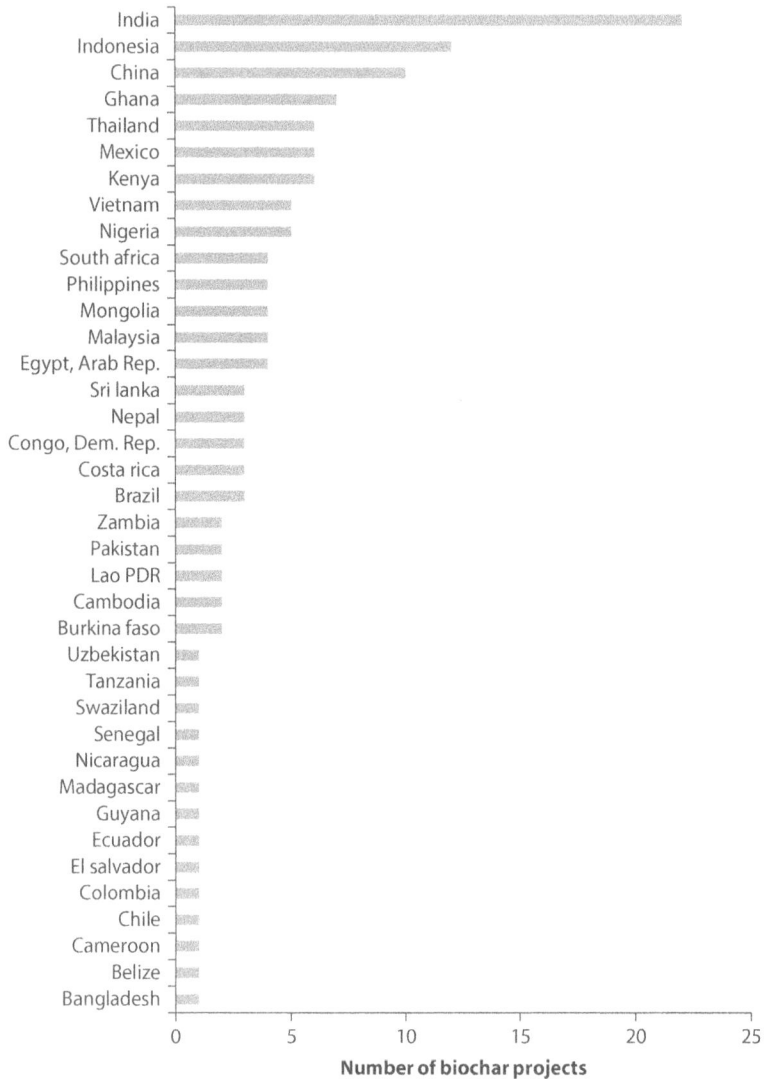

Source: World Bank.
Note: While the primary aim of the survey was to shed light on biochar projects in developing countries, a number
of responses were received from developed-country projects and are included in this figure and in the data
analysis below.

second survey are discussed in chapter 6. Results of both the initial surveys are
presented in appendix A.

Classification of Biochar Systems

Typology Design

As biochar systems move from the drawing board to implementation, concepts
are tested and systems either advance toward adoption or are abandoned. Even
in the process of setting up pilot projects, ideas that may have seemed feasible

initially are often modified as the realities at a particular location become clearer. The survey of the IBI network is, necessarily, a snapshot in time of a rapidly growing and changing field of endeavor. Even so, some clear trends can already be perceived in the types of systems that are being proposed and implemented, particularly in developing countries. This section discusses those trends and constructs a preliminary typology of systems using the survey data. This typology is based on production technology, energy use, feedstock choice, and project scale, as outlined in the following subsections.

Production Technology

There are many possible ways to classify biochar systems. One important consideration is the type of biochar production technology that is used. Most biochar production technologies represent a significant capital investment and once the technology choice is made, it may be difficult to change later. Figure 4.2 shows the distribution of different production technologies that survey respondents reported. The choice of technology is closely tied to the available feedstocks—most feedstocks perform best in the particular technologies most suited to handle their characteristics, whether those are chemical composition, size and shape, or degree of moisture. However, it is also the case that many of the production technologies are flexible enough to handle a variety of feedstock types.

Many of the biochar survey respondents (32 percent) make biochar in traditional charcoal pits or mounds that are known to be both polluting and inefficient. However, this is often a way to initiate testing biochar in their soils using

Figure 4.2 Biochar Production Technologies

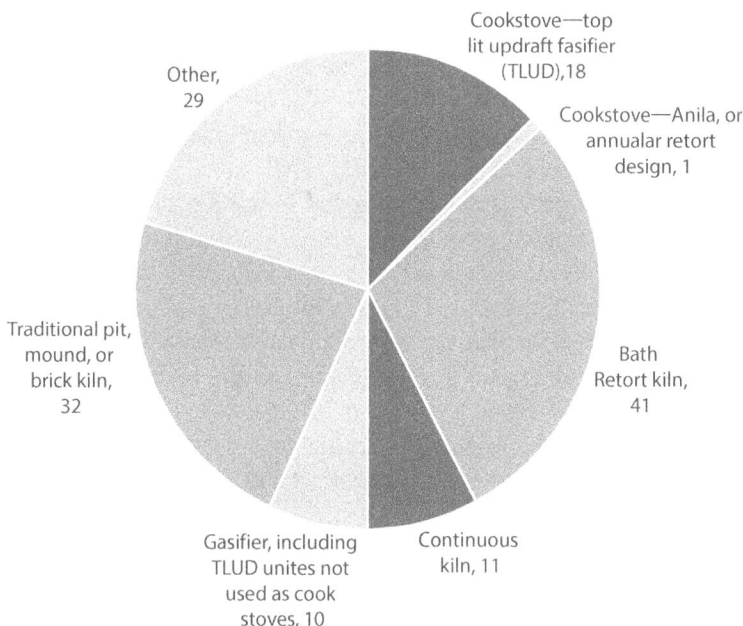

Source: World Bank.

existing technology to see if there is enough benefit to justify investing in cleaner, more efficient biochar-producing technology. In some cases, there is no active technology choice because the biochar projects are utilizing charcoal fines from existing fuel charcoal production. If charcoal is being produced currently for fuel, there are strong reasons to deploy cleaner, more efficient technologies such as batch retort kilns (41 percent) and continuous kilns (11 percent), no matter what the end use of the charcoal is.

Several different kinds of biochar-making stoves exist, and new ones are being developed, with two major types being represented in this survey. The top-lit updraft (TLUD) gasifier is the most common stove design represented here (18 percent of total), while the Anila-style stove uses an annular retort design, and is further described in chapter 5.

Another relatively prevalent technology (10 percent) is the gasifier (not used as a cookstove). Most gasifiers are designed to gasify all the biomass carbon, including most of the char. Gasifiers that generate electricity and that also produce significant amounts of biochar are usually rice husk gasifiers. Rice husk gasifiers currently produce biochar only as a by-product, but the volume of this carbonized rice husk can be quite large because the high silica content of the rice husk inhibits gasification of the char. An inherent inefficiency in the gasification process thus becomes a benefit for those seeking a source of biochar.

Energy Use

Technology choices are also influenced by the energy recovery that is desired. Only 51% of the projects surveyed indicated that they were or would be capturing useful energy released during biochar production. Of these, cooking was the largest single energy use (figure 4.3). Heat energy is the easiest kind of energy to recover from a pyrolytic gasifier, but the small size of cookstoves limits the amount of biochar produced. Several projects indicated in comments that they were including both biochar-making cookstoves and a larger batch or continuous kiln dedicated to biochar production in order to produce adequate supplies of biochar for their specific applications.

While a large number of projects were interested in using recovered heat for drying crops or produce, relatively few projects set their sights on generating electricity. There were two basic approaches to generating electricity used in the surveyed projects: generators driven by internal combustion engines running on gas supplied by a thermal biomass gasifier, and solid-state thermoelectric generators attached to stoves operating on heat conducted through the metal stove body. Thermoelectric generators are low-output devices that are fairly costly. They are suitable for tasks such as cell phone charging and running LED lights. Large biomass gasifiers are capable of generating several to hundreds of kilowatts of electrical power.

Feedstock Choice

Another important criterion in classifying biochar systems is feedstock use. For example, some stoves can use unconsolidated crop wastes such as straw, but the

Figure 4.3 Utilization of Biochar Production Energy

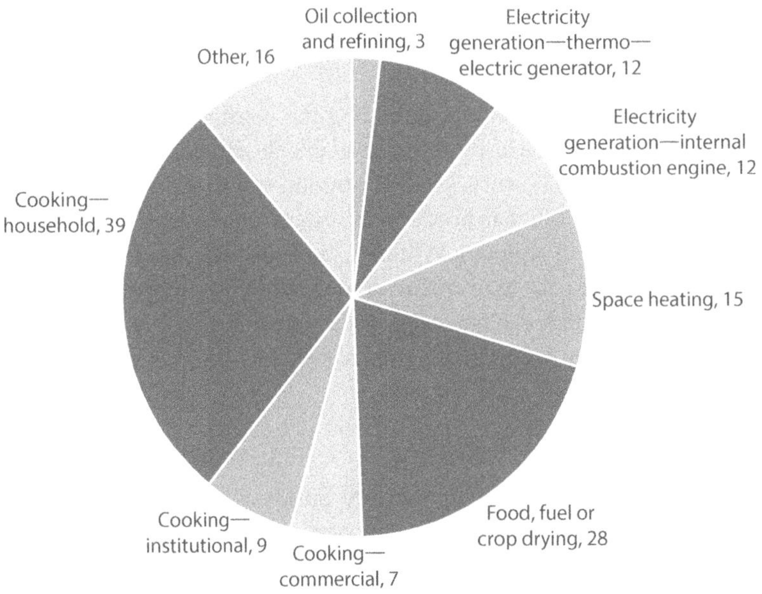

Oil collection and refining, 3

Electricity generation—thermo—electric generator, 12

Other, 16

Electricity generation—internal combustion engine, 12

Cooking—household, 39

Space heating, 15

Cooking—institutional, 9

Cooking—commercial, 7

Food, fuel or crop drying, 28

Source: World Bank.

Figure 4.4 Word Cloud Showing Biochar Feedstocks Most Frequently Cited by Survey Respondents

Source: World Bank.

most common design, the TLUD stove, requires either small feedstocks, such as nut shells or seeds, or small briquettes or pellets that can be made from a variety of crop wastes or weeds. The survey used an open-ended question format to collect information about feedstocks. This returned a wide range of results that was very useful for the purpose of sampling the spectrum of potential feedstocks, but was rather difficult to classify. Figure 4.4 displays a visual impression of the variety of feedstocks used and the frequency of references to them (word size represents frequency of use).

Project Scale
The project scale determines key factors, including the community that will be influenced by the project, the area of potential available feedstocks, and the types of energy that may be useful. The survey asked respondents to answer the question, "What is the scale of the biochar production system?" choosing from a scale that ranged from household to regional. While it was expected that the answers would roughly correspond to the size of the actual pyrolysis technology, there may be some discrepancies, such as a far-reaching cookstove project. As figure 4.5 shows, the majority of biochar production systems were identified as small, tending toward household- and farm-scale systems.

Typology Construction
To construct a typology of biochar systems based on the survey response data, each project was categorized based on energy recovery type, scale, feedstock type, and production technology. For the scale categories, *village* and *cooperative* scales were combined to simplify the presentation. Because many different feedstocks were used, six categories were designated. Feedstocks such as field stover, nut hulls, empty palm fruit bunch, and bagasse were all grouped into the category of *crop residues*. Because there were a large number of references to rice crop

Figure 4.5 Scale of Biochar Production Systems

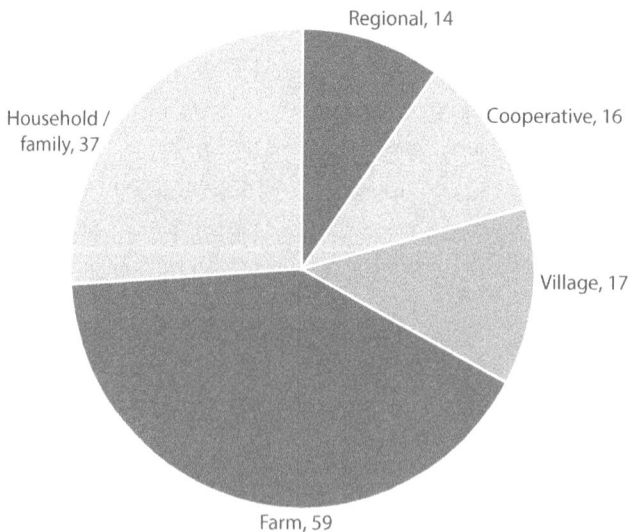

Source: World Bank.
Note: Due to the nature of the survey, the breakdown above should not be considered representative of all biochar projects.

residues (mostly rice husk), *rice residues* was kept as a separate category. *Wood residues* include sawmill residues, fuelwood, urban greenwaste, and plantation thinning and prunings. The *manure* category is mostly composed of dried dung. *Weeds* are invasive species, both herbaceous and woody, with water hyacinth and *Prosopis juliflora* being the primary examples. *Slash and char* refer to projects that are attempting to slow down cycles of slash-and-burn agriculture by returning char rather than just ash to newly cleared fields in the hope that they will retain productivity for longer. The *other* category includes several projects that were using charcoal fines from an existing charcoal production process (for fuel), as well as all of those projects that declined to state what feedstocks they were using.

After categorizing all the projects, they were plotted to make the bubble chart in figure 4.6, where the size of the bubble indicates the number of projects for each feedstock type at a given combination of energy type and scale. Total bubble size is not additive—bubbles lie on top of each other. The differently shaded bubbles represent different feedstocks, and these are plotted with energy on the *y*-axis and scale on the *x*-axis.

The first aspect to note from figure 4.6 is that cooking energy predominates at the household scale, as would be expected. Also, most of the projects

Figure 4.6 Typology of Biochar Systems by Type of Energy Recovery and Scale Showing Number of Projects with Each Type of Feedstock (*n* = 154)

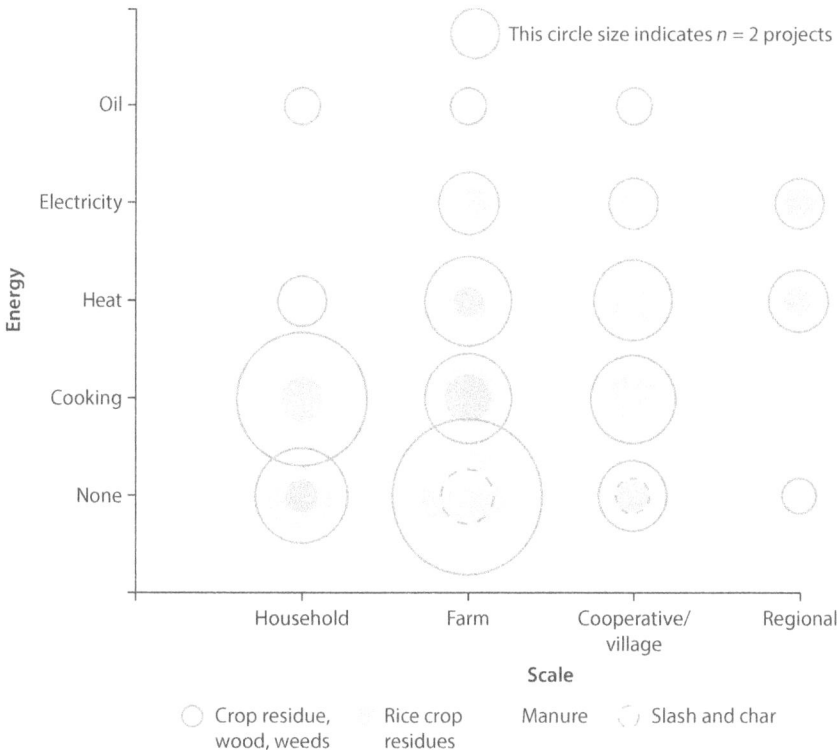

Source: World Bank.

generating electricity are at the larger scales. The farm scale is the most likely to be producing biochar without any energy capture, indicating that the system may be driven primarily by the agronomic benefits of biochar. As the complexity of the form of energy increases (toward oil and electricity), the number of projects decreases.

Crop residues, wood residues, and weeds were used under similar conditions (figures A.11 and A.12 in appendix A), so they were grouped together in figure 4.6, and join rice residues to be the predominant feedstocks. Rice residues, however, were primarily used at the farm or cooperative/village scale, rather than household or regional scales. The slash-and-char systems surveyed had no energy recovery and were concentrated at the farm or cooperative/village scale. These trends are summarized in figure 4.7.

These typology charts reflect all of the 154 project survey responses. Keeping in mind that more than 40 percent of the projects identified themselves as phase 1 projects (assembling concepts, partners, and funding), a table of system typologies was constructed, grouping projects hierarchically based on the most common types and the level of similarity between projects (table 4.1). The relative importance of each category was determined using the dendrogram in figure A.13 in appendix A, which demonstrates that the most influential grouping category is scale, followed by energy use, and feedstock. The common production technologies used for each of these is also listed, although they are, of course, closely tied to the energy use in many cases.

The work of developing biochar system typologies is in the early stages and will evolve as biochar projects advance. Once the available feedstock opportunities are identified, biochar production technologies should be assessed and developed that are suitable to the scale and energy needs of the farmers and project participants. In addition, consideration will need to be given to all of the other social, ecological, and economic components of the agricultural and energy systems in which biochar may play a role.

Figure 4.7 Summary of Dominant Biochar Typologies

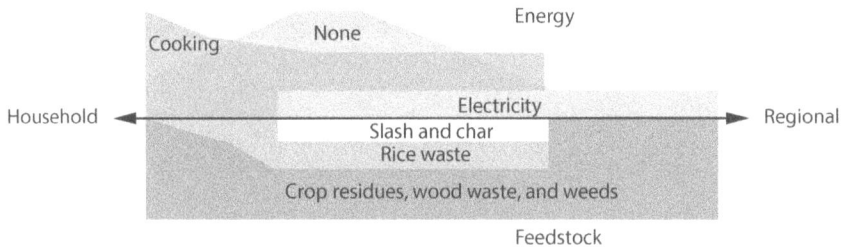

Source: World Bank.
Note: Arrow designates scale, from household, to farm, to cooperative/village, to regional, while areas above the arrow designate prevalence of energy uses and areas below the arrow designate prevalence of feedstock uses. Energy uses and feedstocks without strong trends are not indicated on this figure; areas are giving only a relative approximation of importance.

Table 4.1 Biochar System Typology

Systems	Energy use	Feedstock	Production technologies[a]	Category
Household	**Cooking**	All sources	cookstove, TLUD; cookstove, Anila or annular retort design; batch retort kiln; traditional kiln	Ia
	None	All sources	batch retort kiln; traditional kiln	Ib
	Heat	Crop/wood/weed	cookstove, TLUD; batch retort kiln	Ic
	Electricity	Wood	batch retort kiln	Id
Farm	**None**	All sources but manure; *slash and char*	traditional kiln; batch retort kiln	IIb
	Heat	All sources except for rice residues	traditional kiln	IIc
	Electricity	Crop/wood/rice	gasifier; batch retort kiln	IId
	Oil	Crop/wood	commercial pyrolysis pilot plant	IIe
Cooperative/ village	Cooking	Crop/wood/weed/*rice*	gasifier; traditional kiln; batch retort kiln	IIIa
	None	All sources; *slash and char*	batch retort kiln; continuous kiln	IIIb
	Heat	Crop/wood/weed/*rice*	continuous kiln; batch retort kiln; traditional kiln	IIIc
	Electricity	*Rice*/weeds	gasifier	IIId
	Oil	Crop	gasifier	IIIe
Regional	None	Crop/wood	continuous kiln	IVb
	Heat	Crop/wood/manure/ weeds	gasifier; continuous kiln; batch retort kiln; traditional kiln	IVc
	Electricity	Crop/wood/manure	continuous kiln; batch retort kiln	IVd

Source: World Bank.

Note: The most prevalent categories of energy use at a given scale, as seen in figure 4.6 and figures A.11–A.13 (appendix A), are shown in **bold**, while the categories in which a given feedstock is concentrated are shown by *italicizing* the feedstock.

Categories are I for household, II for farm, III for cooperative/village, and IV for regional scale biochar system. The letter (a) categorizes energy as being used for cooking, (b) no energy use, (c) the off heat is being used, (d) the biochar system produces electricity, (e) the system produces bio-oil.

a. Note on terminology: TLUD = top-lit updraft gasifier; "traditional" kilns were described in the survey as "traditional pit, mound or brick kiln."

CHAPTER 5

Life-Cycle Assessment of Existing Biochar Systems

Life-Cycle Assessment: Definition and Methodology

Life-cycle assessment (LCA) is a methodology to evaluate the environmental flows associated with a product, process, or activity throughout its full life by quantifying energy and resources used and emissions created. The life cycle of a product "from cradle to grave" includes four principal phases: materials extraction and processing, product manufacture, product use, and product disposal or "end of life." The LCA methodology used in this study has been standardized by the International Organization for Standardization according to the 14040 series (ISO 1997). Because of its whole systems approach and transparent methodology, LCA is an appropriate analysis for estimating the global warming impacts of biochar systems. In addition to its rigor as an environmental impact assessment tool, economic costs can be included in an LCA, making it a versatile method for analyzing the profitability of a given system.

In order to avoid unintended consequences it is necessary to conduct analyses of potential life-cycle impacts of biochar systems. It would be undesirable to have the system actually emit more greenhouse gases (GHGs) than are sequestered or consume more energy than is generated. Analyses such as LCA provide not only insightful information as to the potential impacts of biochar systems, but also a framework for considering the production chain of feedstocks and transparency in defining system boundaries for the assessment. In addition, an LCA can also be used as a tool to identify data gaps and key processes or effects within the life cycle that have the greatest impact on the results. This information can be used to direct future research and system improvements. The four main elements to the LCA methodology are (a) a goal and scope definition; (b) the inventory analysis; (c) the impact assessment; and (d) the interpretation, as described in box 5.1.

The contribution analysis calculates the relative impact of different life-cycle stages (for example feedstock production, transportation, pyrolysis) to identify those stages with the most impact, and to identify opportunities to reduce GHG emissions and costs of the system. The sensitivity analysis determines how the results may be affected by uncertainty or variability in the data and between systems.

Box 5.1 Elements of a Life-Cycle Assessment

Goal and scope definition
- Clearly defining the breadth of the study, the system cutoffs, and the system boundaries

Inventory analysis (life-cycle inventory)
- Researching and collecting the process data with all of the input and output flows of the system
- Developing the life-cycle inventory model

Impact assessment
- Analyzing the environmental consequences of the flows of the inventory analysis

Interpretation
- Contribution analysis
- Sensitivity analysis
- Data quality assessment
- Recommendations for pollution prevention and resource conservation.

Goal and Scope

The purpose of the LCA case studies is to conduct cradle-to-grave analyses of the GHG and economic inputs and outputs of biochar systems in developing countries.

System Boundaries

The system boundaries are best defined and illustrated in a process flow diagram for each case study. Determining the system boundaries in LCAs is important for defining allocation procedures. Allocation issues arise when a multioutput process is defined in terms of one product. The question arises how to separate or designate the impacts of only one of those products. The ISO guidelines state that whenever possible allocation should be avoided by (a) increasing the level of detail of the model; or (b) using the system expansion method. Although increasing the level of detail is preferred, it is not always possible because of data limitations. System expansion and allocation by partitioning are methods for handling the allocation problem. System expansion means that the system boundaries include all processes affected by changes in the studied system. System expansion includes the avoided processes as part of the life cycle of the product, that is, the avoided processes are subtracted from the model. Allocation by partitioning is done when the impacts are divided among the coproducts, based on weight, energy content, economic value, or similar (Baumann and Tillman 2004). The case studies utilize both methods as deemed necessary and are described in each study separately.

Impact Categories

The impact categories for this assessment are the global warming impact (total GHG emissions and reductions) and the net economic impact (total costs and revenues).

Input/Output Data Requirements

The cutoff criteria for the input and output data have been set as follows: if any given input or output is less than 5 percent of the energy or mass of the feedstock per functional unit, these data are not included if they are not available. If data are available, they are included regardless. Any inputs or outputs that are not included are noted in the process descriptions.

Data Quality Assessment

The data quality requirements are those outlined in the ISO 14040 standards and Ansems and Ligthart 2002, in addition to the methodology developed by the University of Washington's Design for Environment Laboratory. The data quality assessment utilizes a scoring method based on time, geographic, and technology coverage parameters, and precision, completeness, representativeness, consistency, and reproducibility considerations. Each process is assigned a data quality score for each indicator on a scale of 1 to 5, with 1 being the best. The assignment of a score to each category is based on the capability of the dataset to meet specific requirements. The average of the scores was calculated for each process. The results from the data quality assessment can be found in appendix B.

Life-Cycle Inventory

Detailed descriptions of the subprocesses and their input/output data are included under each case study, using an adapted version of the biochar energy, greenhouse gases, and economic (BEGGE) model (Roberts et al. 2010).[1]

Impact Assessment

The common 100-year global warming potentials of carbon dioxide, methane, and nitrous oxide (1, 25, and 298 CO_2e, respectively) from Intergovernmental Panel on Climate Change (IPCC) 2007 were used to calculate the climate change impacts of each process. The net climate change impact is the sum of the net GHG reductions and the net GHG emissions. To be consistent with terminology, the "net GHG reductions" are the sum of the "CO_2e sequestered" and the "avoided CO_2e emissions." The carbon sequestration is a direct result of the stable carbon in the biochar, while the avoided emissions are from the avoided processes such as traditional cooking and crop residue burning. The biogenic carbon dioxide is fully accounted for in the analysis in the following manner: the carbon uptake during feedstock growth is included for sustainable feedstock harvest but not for unsustainable feedstock, and the carbon emissions during thermal conversion and the carbon sequestration in the biochar. None of the case studies include improvements to the soil structure, or direct impacts on BC settlement in the cryosphere due to lack of data. These differences or absence of specific soil improvements could further impact GHG emissions, some of which are explored in the sensitivity analyses for each case study.

The costs and revenues of the biochar systems are dependent on the nature and organization of each biochar project. All monetary costs are included as appropriate (for example cookstove, transportation) as compared to current practices. The changes in time or labor (collecting feedstock, applying biochar) of the biochar system compared to the traditional system are monetized where possible, and when not possible they are tracked separately from the economic costs and revenues and then monetized as part of the discussion. Carbon offset credits are currently not received by any of the case study projects, thus are only included in the analysis as a discussion item.

Case Studies

Criteria for Selection

The initial survey of the International Biochar Initiative (IBI) network (a global network of biochar practitioners that includes academic researchers, commercial interests, farmers and gardeners, government representatives, and nongovern- mental organization project developers) to identify biochar projects in develop- ing countries yielded preliminary data on roughly 150 different projects in 41 countries, as described in chapter 4. About half of these projects are still in the planning stage. Therefore, the initial selection filter was to identify projects that were actually being implemented on the ground and likely to have reportable results on at least some of the key parameters examined in the LCA (for example feedstock characteristics, crop yield, energy use). Looking for only the most com- plete project datasets, the selection was narrowed down to 25 projects. From that group the top three projects were selected using a more complete set of criteria, which are described further below (in no order of priority):

1. **Degree of integration into local economy.** A project having connections to the local economy is more representative than a research project conducted by academics only. Although several academic research projects had good crop yield data, they lacked any integration into the surrounding real-life econom- ic setting, meaning they had few or no data about labor and other inputs and how these projects would actually function and operate in the "real world." However, some academic research projects are working directly with farmers and in many ways these are useful cases because they tend to have high- quality data.
2. **Variation in biochar production technology and scale.** The case study selec- tion was oriented toward smallholders and cooperatives rather than industrial scale. The first selected Kenya project is a cookstove project, so the selection process focused on small batch kilns and medium-scale pyrolyzers or gasifiers as alternative biochar production methods.
3. **Data availability and quality.** None of the projects submitted had an entirely complete dataset for the LCA that included, for instance, multiyear crop yield data measured scientifically against a control. In every case, at least some pa- rameters are considered primarily via a sensitivity analysis.
4. **Replicability in a broad range of contexts.** The projects were screened for replicability to ensure that those analyzed in the LCA would provide poten-

tially wide-ranging insights if the models showed that they were successful in reducing GHG emissions or increasing agricultural production, flexibility, and climate resilience. Variables that guided the selection process included the availability of waste material as feedstock, soil conditions suitable for biochar application, and biochar production technology that could be economically fabricated and used in different, predominantly rural settings.

5. **Impact on GHG reduction (climate change mitigation effect).** Admittedly, prior to the LCA it was not entirely certain whether or not a project would have a positive climate change mitigation effect. However, key input parameters such as feedstock uses and transportation were considered to eliminate any obviously unsustainable projects.

6. **Geographic location.** The goal of the study was also to look for a variety of project locations in low-income developing countries where the need is greatest for climate-smart farming practices that improve rural livelihoods while mitigating and adapting to climate change.

Project Descriptions

Ultimately, three projects in different geographic locations—Kenya, Vietnam, and Senegal—were selected for the LCA based on the above criteria. The Kenya case study is based on a household-scale pyrolysis cookstove that produces biochar in addition to providing heat for cooking for subsistence farmers. The Vietnam case study represents biochar produced as a by-product of small-scale rice wafer production from rice husk feedstock, and the biochar is purchased by local peanut farmers. The Senegal case study looks at a larger-scale continuous process pyrolysis unit that produces biochar and is used by local onion farmers. In-depth descriptions of each case study are provided below.

Kenya Case Study Life-Cycle Assessment

System Overview

The project in Kenya is conducted by Cornell University, under Professor Johannes Lehmann of the Department of Crop and Soil Sciences. Improved pyrolysis cookstoves are being tested in subsistence farming households in the western part of Kenya, where biochar and heat energy for cooking are coproducts of utilizing these stoves. The biochar stove under consideration uses primary (woody) and secondary (herbaceous) feedstocks for operation. Crop residues from the farm that are not otherwise used (for animal feed or other household needs) can be used as the secondary feedstock as well, thus decreasing primary fuelwood consumption and deforestation pressures. The case study assumes the biochar is applied to maize fields on the farm. However, the biochar could also be applied to vegetable gardens, but the agronomic trials (and thus data availability) have focused on maize crops to date. The biochar trials were begun in 2005, and the pyrolysis cookstove testing began in 2008. The project involves approximately 60 farms in the region.

Currently, the pyrolysis cookstove in this project is in the prototype stage and has not been optimized yet. The pyrolysis cookstove has been brought to households for testing for a week at a time, but families have not had extended use of the stove so far. An improved design of the pyrolysis cookstove is in development. The biochar produced using the cookstove is not currently applied to the family's crop fields, but rather is used for testing. The crop yield data are from farms in the same region, which are sometimes the same households as the ones for the biomass assessment and cookstove measurements. However, the biochar in the field trials was produced not through cookstoves, but in a kiln from woody biomass feedstock (more detail in "Biochar Effect on Maize Yield" subsection below), and applied at a rate of 18 tonnes of carbon per hectare, or 27 tonnes of biochar per hectare. The goal of the project is for each household to have an improved pyrolysis cookstove and apply the biochar to their own fields at an application rate that is high enough to see agronomic returns (actual biochar application rates might well turn out to be much lower than the 18 tonnes of carbon per hectare used in the field trials to date).

Methodology
Function, Functional Unit, and Reference Flows
Biochar production is a multioutput system with up to four coproducts for the Kenya cookstove system: biomass management, soil amendment, carbon sequestration, and bioenergy production. The outputs of the pyrolysis process are dependent on the needs of the producer, and the relative amounts of the final products can be altered by feedstock selection and processing conditions. The functional unit of the pyrolysis cookstove system is 1 tonne of dry crop residues, used as secondary feedstock for a household pyrolysis cookstove on a farm in Kenya. Results are also presented with alternative functional units (such as the yearly cooking energy for one year per household, and 1 tonne of biochar), taking into consideration the cookstove technology and biochar production limitations, as well as the biomass availability limitations on the farm, as the quantity of biochar produced is in fact limited by the cookstove and the biomass.

System Boundaries
A flow diagram of the biochar system is illustrated in figure 5.1. The system is organized into four modules (bold outlined boxes): feedstock, cookstove, biochar, and crop response. Under each module a summary of subprocesses is listed.

Scenarios
Throughout the analysis, the so-called "baseline" scenario is the most reasonable pyrolysis cookstove scenario, which is not to be confused with a "business as usual" scenario. The business as usual scenario is the current traditional practice (cooking with a three-stone fire), and is actually included in the baseline scenario by subtracting out these avoided traditional practices in the LCA model (see detailed process descriptions in the following "Life-cycle Inventory" section).

Figure 5.1 Schematic Flow Diagram for Biochar Production in a Pyrolysis Cookstove System

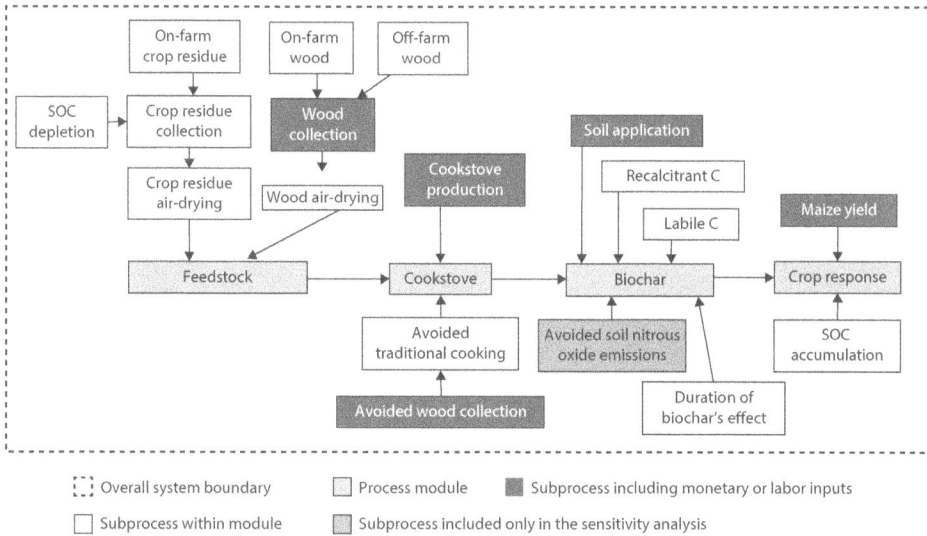

Source: World Bank.
Note: SOC = soil organic compound.

The pyrolysis cookstove for this case study is a rocket-style Anila stove (figure 5.2), which utilizes primary and secondary feedstocks. The Anila stove operates with a rocket design, where the primary feedstock (woody biomass) is inserted through an opening on the lower part of the stove and is combusted in the center chamber of the stove. The outer ring of the stove is the pyrolysis chamber, which utilizes the secondary feedstock and can accept herbaceous biomass such as crop residues. The outer chamber is a low-oxygen environment, and the gases driven off the secondary feedstock flow out of holes to the primary chamber where they are combusted, thus adding to the cooking energy of the primary fuel. In comparison, a traditional three-stone set-up is an open fire using woody biomass as the only feedstock, where usually three stones or bricks are set around the fire and the cooking pot is set on top of the stones.

Because the crop residue feedstock is not purposefully grown, the production of the feedstock is considered a by-product and therefore no environmental burdens are associated with its generation (except for those impacts that would not otherwise occur in conventional management). The feedstock is collected, air-dried, and chopped to fit the cookstove as necessary for specific feedstock types. The fuelwood for primary combustion in the pyrolysis cookstove is included in this module. The amount of primary fuelwood required for the pyrolysis cookstove is less than for a traditional cookstove. Thus, the difference in primary fuelwood collection is accounted for as an avoided process in the feedstock module. Without a pyrolysis cookstove, these same crop residues would be left to decompose in the field (those not collected for animal feed or other household uses that are considered not available for cooking).

Under the cookstove module, the production and transportation of the cookstove are accounted for. During the cookstove operation, slow pyrolysis, cooking

Figure 5.2 Pyrolysis Cookstove in Kenya Case Study

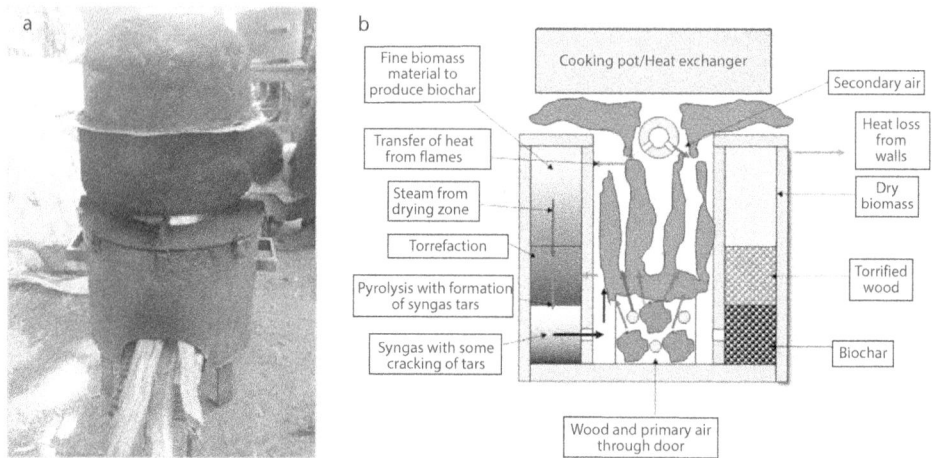

Sources:
5.2a. Prototype pyrolysis cookstove in use in Kenya (image courtesy of D. Torres, 2010).
5.2b. Schematic diagram of the rocket-style Anila pyrolysis cookstove (image courtesy of S. Joseph, International Biochar Initiative).

energy, and air emissions are included. The avoided emissions associated with a traditional cookstove are part of this module. The biochar is transported and applied to soils by hand. The behavior of biochar in soils is described by the stability of the carbon in the biochar, where the recalcitrant and labile carbon fractions are included in the biochar module. The crop response upon biochar application is compared to yields for control crops. The effect of biochar on soil nitrous oxide emissions is a subprocess illustrated in the flow diagram that is not in the baseline scenario but is included in the sensitivity analysis. The system expansion method is used for modeling avoided biomass fuel collection, avoided biomass management, and avoided traditional cookstove cooking.

Life-Cycle Inventory
Feedstock
Torres et al. (2011) conducted a detailed biomass assessment of farm households in the study region. In this work the total biomass was estimated for 50 households. The available feedstock for pyrolysis was calculated by subtracting the biomass required for other household and farming needs, such as animal fodder and building materials, from the total biomass. Thus, in this LCA both primary (wood) and secondary (crop residue) biomass collected on the farm are "sustainable" in that they account for other biomass requirements and harvest practices that allow regeneration of the biomass. (However, this does not account for soil cover against erosion, as discussed later.) Based on these data, the average on-farm wood available for cooking is 1.62 tonnes of dry matter per household per year, and the on-farm crop residues available are 3.33 tonnes of dry matter per household per year. From table 5.1 it is evident that the primary and secondary feedstock consumption for the pyrolysis cookstove is met with sustainable

Table 5.1 Primary and Secondary Feedstock Characteristics and Availability for Baseline Scenario

	Wood	Residues	Unit
Moisture content	16.28%	68.00%	Dry matter on wet basis (70°C for 48 hours)
Carbon content	47.86%	46.29%	
On-farm availability[a]	1.62	3.33	Tonnes per household per year
On-farm pyrolysis feedstock consumed	1.15	1.87	Tonnes per household per year
Off-farm pyrolysis feedstock consumed	0	0	Tonnes per household per year

Source: World Bank.
a. Based on sustainable harvest and other biomass uses.

on-farm supplies. The on-farm wood comes from woodlots composed mostly of fast-growing single species of trees and from trees and shrubs scattered throughout the farm and around the perimeter (Torres et al. 2011). Crop residues available on the farm are predominantly maize stover, in addition to collard green stalks and banana leaves. Table 5.1 lists the primary and secondary feedstock properties, availability, and quantities consumed using a pyrolysis cookstove per household per year. As the fraction of available banana leaves and collard green stalks are only 5 percent and 0.6 percent respectively of the total available residue, the secondary feedstock characteristics utilize maize stover data. However, the other crop residues would not significantly affect the results, as the feedstock properties (heating value and carbon content) are similar to that of stover (Phyllis 2008).

The on-farm wood collection is assumed to occur regardless of the household having a traditional or pyrolysis cookstove because the wood fuel requirements for a traditional stove would consume all on-farm wood, plus off-farm wood. Thus, time for collecting on-farm wood is not included. The off-farm wood collection requires significant time, and was determined from a self-reported survey of 31 households in the region, where the time to collect off-farm fuelwood was estimated at 4.6 minutes per kilogram of dry wood.[2]

The crop residues are collected entirely from the farm under the baseline scenario. During maize harvest, it is common practice for farms larger than about 1 hectare to cut the entire aboveground maize plant and make a windrow of plants to allow the grain to dry.[3] After drying the grain is removed. Approximately 25 percent of the stover is used for animal feed or other purposes (Torres et al. 2011), and the remaining stover is typically applied back to the field.[4] Farms smaller than 1 hectare are more likely to harvest each individual maize, and cut down the stalk and leave it in the field.[5] Those residues not used as fodder or fuel decompose in the field. As the average farm size from the study region is 1.77 hectares, the baseline scenario assumes the stover residue is collected regardless of its use in the cookstove. Therefore, the time to collect on-farm residue is not included in the LCA.

As 25 percent of the total aboveground crop residue is utilized for fodder, construction, or other household purposes, it is assumed that the remaining 75 percent of the residue would be returned to the fields if a traditional three-stone cookstove were in use. Returning crop residues to the field is important for minimizing erosion. The amount of residue retained on the field to minimize erosion is dependent on soil, slope, climate, and crop. Under the baseline scenario, all secondary feedstock (crop residue) needed for the pyrolysis cookstove is assumed to be collected on the farm. Based on the pyrolysis cookstove's rate of secondary fuel consumption and measurements of on-farm biomass, 42 percent of the total aboveground crop residues are collected as secondary fuel for cooking and biochar production, while 33 percent of the total aboveground crop residues are returned to the soil, in addition to the biochar that is applied to the soil. The total average aboveground residue available per household is 4.44 tonnes of dry matter per household per year (2.5 tonnes of dry matter per hectare, using the average farm size of 1.77 hectares). Thus, 1.46 tonnes of dry matter per household per year (0.82 tonnes per hectare) of residue would be returned to the soil, 1.11 tonnes used for fodder, and 1.87 tonnes available for pyrolysis. The 33 percent crop residue returned to the soil may or may not be sufficient for erosion protection, as studies specific to the region have not yet been conducted. Limited crop residue production and competing uses for residues on the farm are challenges for soil conservation (Unger et al. 1991). A seven-year study of the effect of mulch rates and tillage systems on oxisols in Brazil recommends that 4–6 tonnes per hectare of residues be kept on the soil to reduce runoff and erosion most effectively (Roth et al. 1988). The study found that the soil cover must be at least 90 percent for complete infiltration of high rainfall amounts, where the typical quantities of 1.5 and 2.5 tonnes per hectare of wheat and soybean residue left on the field, respectively, covered 60 percent of the soil surface. From the Kenya case study, even if all the residues were left on the field, 2.5 tonnes of dry matter per hectare would be insufficient to fully prevent runoff and erosion, according to Roth et al. (1988). Given the limited residue production and competing uses, additional strategies for erosion protection, such as selective residue removal, substituting high-quality forages for residues, alley cropping, and using wasteland areas to grow biomass (Roth et al. 1988), should be considered. If the 33 percent residue return is insufficient for erosion protection, then some of the secondary feedstock for the pyrolysis cookstove would need to be collected off the farm. For example, if 50 percent of the total residues were retained for erosion prevention, an additional 0.76 tonnes of dry matter of off-farm residue would be needed each year. Sources of herbaceous residues could come from local sawmills or roadside vegetation. The effects of retaining all residues on the field and collecting secondary feedstock off the farm are discussed in the sensitivity analysis.

Cookstove Construction

The fuel use, cooking, and biochar production data are from direct measurements using a prototype pyrolysis cookstove made of metal. The stove was built at Cornell University and brought to Kenya for testing. It is anticipated that the cookstove design could be duplicated using scrap metal locally available in Kenya, or from clay, as is currently under way in the project. The mass of the stove is approximately 2.25 kilograms. The cost of the stove to the farmer is estimated at $10, based on the price of other scrap metal stoves designed to burn sawdust. The lifetime of cookstoves is dependent on the material, construction, and use of the stove. The stoves in this project are still in the early stage and have not come to the end of their useable lifetime. Based on the literature, a range of three to five years is typical for improved biomass cookstoves (Johnson et al. 2008; Joseph 2009; Qadir and Kandpal 1995). The baseline scenario assumes a stove lifetime of four years.

Stove Transport
It is assumed that the stove is produced at the market center. A farmer would typically walk from home to the village center and then take a bus from the village center to the market center. The distance from the village to the market center is 10 kilometers each way. The cost of the bus for one passenger is 30 Kenyan shillings each way, or $0.36. The GHG emissions and cost of the round trip bus transport for the farmer to get the stove and bring it back to the farm are included. Emissions data from buses in Kenya were not available, therefore emissions from an urban diesel bus in the United States are used from Sheehan et al. 1998. It is possible that the emissions from a bus in Kenya are higher than from a U.S. bus; however, the overall contribution of the bus emissions to the life cycle are less than 1 percent of the total GHG balance, below the cutoff requirement.

Pyrolysis and Cooking
The amount of feedstock consumed throughout the year by the pyrolysis cookstove is presented in table 5.1 based on data available for in-field fuel consumption during household cooking. Two different types of tests were utilized for measuring fuel consumption. Table 5.1 presents one method, assessing energy consumed during daily cooking tests in a sample of 20 households (Torres et al. 2011), which is used as the baseline scenario. Fuelwood, biomass, wood char residues, and biochar were measured during daily cooking activities on a per household basis. To determine the energy used per capita, the amount of fuelwood used is multiplied by the energy content and divided by the total amount of people in the household (average household size is 6.7 people). The resulting energy consumption per capita serves as a baseline to compare the current energy consumption of the household and the consumption when a pyrolysis stove is introduced. The result is the fuel use value in table 5.1 used for the baseline scenario.

The second method for measuring fuel consumption used a kitchen performance test in 17 homes (Whitman, Scholz, and Lehmann 2010), where the mass of fuel in a pile before cooking and the mass of fuel remaining in the pile after a predetermined amount of time for a traditional three-stone cookstove was measured and determined to be 1.95 kilograms of dry fuel per capita per day. Subsequently, a ratio from the first method for primary to secondary fuel use for the pyrolysis stove was multiplied by the daily wood use, yielding the high fuel use value as discussed in the sensitivity analysis. Both fuel consumption measurement methods have strengths and weaknesses, and thus data from both methods are presented.

The yield of biochar is calculated to be 0.52 tonnes per household per year (Torres et al. 2011), which is a 17.2 percent yield of the total feedstock consumption on a dry mass basis, or a 27.8 percent yield for the secondary feedstock only. The yearly biochar yield with this cookstove is higher than the 0.19 tonnes per household per year estimated by Iliffe (2009) for the Anila cookstove.

Emissions Data

Although emissions data specific to the pyrolysis cookstoves is not available, the sensitivity analysis shows that within the tested range the results are reasonable and the same conclusions can be drawn. The emissions are calculated based on data from a gasifier cookstove from MacCarty et al. 2008 and comparing emissions to ratios of carbon dioxide and products of incomplete combustion, using the same methodology as in Whitman et al. 2011. The carbon emitted is calculated from the 0.52 tonnes per household per year biochar yield and a biochar carbon content of 65.89 percent (Torres 2011), at 1.07 tonnes per household per year for the pyrolysis stove (as in table 5.2), and indicates that 24 percent of the carbon is retained in the biochar during pyrolysis and 76 percent is emitted (based on the total carbon in both the primary and secondary feedstocks). If calculated based on the carbon in the secondary feedstock only, then 40 percent of the carbon is retained in the biochar during pyrolysis. The amount of carbon dioxide emitted is calculated from equation (5.1):

$$CO_2 = C_e / [X_{CO_2} + [(CO:CO_2)*X_{CO} + (CH_4:CO)*X_{CH_4}*(CO:CO_2) + (EC:CO)*X_{EC}*(CO:CO_2) + (OC:CO)*X_{OC}*(CO:CO_2)]] \qquad (5.1)$$

where C_e is carbon emitted, X_Y is the molar mass ratio of carbon to compound Y, $CO:CO_2$ is the mass ratio of carbon monoxide to carbon dioxide, EC is elemental carbon, OC is organic carbon, and CH_4 is methane, after Whitman et al. (2011). However, it is important to emphasize that the carbon dioxide emissions from renewable feedstock are cancelled out by uptake of carbon dioxide during regrowth. For the baseline scenario, all feedstock is sourced sustainably on-farm and thus carbon dioxide emissions during pyrolysis of these feedstocks are not included.

Avoided Traditional Cooking

The fuel use for traditional cooking is based on the same biomass assessment by Torres et al. (2011). The total primary feedstock required for traditional cooking is 3.2 tonnes of dry matter per household per year. The available on-farm primary fuel remains at 1.62 tonnes of dry matter per household per year, while the remainder of wood required (1.54 tonnes) is collected off-farm. Off-farm wood

Table 5.2 Calculated Air Emissions from Pyrolysis Cookstove and Avoided Emissions from Three-Stone Fire, per Household per Year

	Pyrolysis	Three-stone (avoided)
Carbon emitted (tonnes per household per year)	1.07	1.51
$CO:CO_2$	0.0252	0.0513
$CH_4:CO$	0.063	0.063
EC:CO	0.00011	0.00011
OC:CO	0.042	0.042
CO_2 (tonnes per household per year)	3.8	2.0
CO (tonnes per household per year)	0.095	0.259
CH_4 (tonnes per household per year)	0.006	0.016
EC (tonnes per household per year)	1.04E-05	2.85E-05
OC (tonnes per household per year)	0.004	0.011

Source: World Bank.
Note: For abbreviations see text accompanying equation (5.1).

can be partitioned into a sustainable or renewable fraction and an unsustainable or nonrenewable fraction. The LCA assumes a nonrenewable fraction of 0.8 for off-farm wood, based on estimates for the high deforestation rates in this region (Müller and Mburu 2009; Whitman et al. 2011). The sensitivity analysis compares varying the fraction of nonrenewable biomass for off-farm wood. The off-farm wood collection requires 117 hours per household per year.

The emissions from the three-stone fire are calculated by the same method as for the pyrolysis stove, as described in the last column of table 5.2. However, in contrast to the pyrolysis cookstove, traditional cooking on a three-stone fire requires more feedstock than can be sourced sustainably (where 80 percent of off-farm wood is nonrenewable), thus the carbon dioxide emissions from the fraction of nonrenewable biomass are counted toward the net GHG impact.

Biochar Soil Application

The 0.52 tonnes of biochar produced annually on the farm is assumed to be applied to maize fields at a relatively high application rate of 18 tonnes of carbon per hectare, as yield data from multiyear field experiments are only available for this application rate in the project. With a carbon content of 65.89 percent in the stover biochar (Torres 2011), this translates to an application rate of 27 tonnes of biochar per hectare. It is assumed that the biochar available is applied at this rate to a small area (0.019 hectares) rather than being spread out over a larger area at a lower application rate. Thus, the effects of biochar on crops and soils are applicable only to this small area of 0.019 hectares per year. Each year sufficient biochar would be produced to apply it at the rate of 18 tonnes of carbon per hectare to an additional 0.019 hectares of cropland. As the average size of the maize plots of farms in the study region is 0.43 hectares (Torres et al. 2011), it would take almost 23 years to apply biochar to the entire maize plot at the rate of 18 tonnes of carbon per hectare. Judging from data of other experiments in acid tropical soil (Lehmann and Rondon 2006; van Zwieten, Kimber, Downie et al. 2010) it is possible that an application rate of less than 18 tonnes of carbon per hectare would be sufficient. Indeed, preliminary results from greenhouse pot trials (Torres 2011) in this region comparing maize growth and nutrient uptake with fresh biomass, biochar, and ash amendments found significant increases in dry matter production (289 percent) at a biochar application rate of around 2.6 tonnes per hectare. However, the crop yield data from the field trials available were based on the 18 tonnes of carbon per hectare application rate, thus that is used as the baseline, and the sensitivity analysis considers lower application rates. Ideally, future research will determine the minimal application rate at which agronomic improvements can be achieved. Another possibility is that the biochar could be applied to kitchen gardens instead of or in addition to maize plots. However, the LCA assumes all biochar is applied to maize because the crop yield data from this region have focused on maize production. The time to apply biochar to the field is based on in-field experience and is estimated at 3.4 hours per tonne of biochar.[6]

Carbon in the Biochar

For 1 tonne of the stover biochar, 658.9 kilograms is carbon. Of this carbon, the majority is in a highly stable state and has a mean residence time of 1,000 years or longer at 10°C mean annual temperature (Cheng et al. 2008; Kuzyakov et al. 2009; Lehmann 2007; Lehmann et al. 2009; Lehmann et al. 2008; Liang et al. 2008; Sombroek et al. 2003; Spokas 2010; Zimmerman 2010). A very conservative estimate is an average half-life of 59 years, as measured across seven sites in the United States using incubation experiments at 30°C (Cheng et al. 2008). The stability of the biochar varies with feedstock type, processing, and environmental conditions. For the baseline analysis, a conservative estimate of 80 percent recalcitrant carbon has been assumed (Baldock and Smernik 2002; Lehmann et al. 2009), and the remaining 20 percent of the carbon is labile and emitted as carbon dioxide in the short term.

Biochar Effect on Maize Yield

Maize crops are grown twice per year in the study region. The maize grain yield data with and without biochar additions are from on-farm field trials from 2005 through 2010 across a chronosequence of years since converting forest to farmland in the same study region and on similar soils, providing the most extensive dataset available. The biochar applied to fields in this dataset is, however, not from pyrolysis cookstoves. The biochar feedstock was *Eucalyptus saligna* Sm. trees, and a traditional kiln method was used with a pyrolysis temperature between 400°C and 500°C. After pyrolysis, the biochar was put into sacks and pounded into small pieces (between 1 and 20 millimeters diameter). The biochar was then spread evenly over the plots and incorporated using a hand hoe (Kimetu et al. 2008). Much recent research has shown that biochar produced from different feedstock and pyrolysis conditions will exhibit different properties (Zimmerman 2010). Thus, it is a legitimate concern that applying crop residue biochar produced in pyrolysis cookstoves instead of the biochar used in the field trials (from eucalyptus trees via a traditional kiln) may result in different soil and crop response. However, the field trial data are the best available at this stage, and the sensitivity analysis will explore this aspect further. Unpublished data from Lehmann (2011) show that stover biochar improves crop yield in the short term more than wood biochar, thus the crop yield data from wood biochar may be a conservative estimate for the period covered in this study.

For this dataset, the biochar was applied at 6 tonnes of carbon per hectare three times over 2005–06 for a total application of 18 tonnes of carbon per hectare, as described by Kimetu et al. (2008). It is important to note that no further biochar applications were made; therefore a total of 18 tonnes of carbon per hectare was applied over the total timeframe of 2005–10. There are two maize crops per year, one each during the long rain and short rain seasons. The long rain yield data for 2005, 2006, and 2008 are based on plots having received the biochar plus full phosphorus and potassium fertilization but no nitrogen fertilization, and controls receiving no biochar but full phosphorus and potassium fertilization (data from 2007 were unavailable). The yield data from 2009 and 2010 were for biochar plots with no fertilization as compared to control plots receiving no fertilization. The motivation behind using plots with minimal or no fertilization is that many smallholder farmers in the region cannot afford to fertilize their plots at the full rate as would be used on research plots. Thus, the best data available with minimal or no fertilization were selected to calculate the baseline. With these data, the average maize grain yield across all farms on control plots is 2.6 tonnes per hectare for the long rain season. Data for the short rain season were only available for 2004, therefore the mean yield across the farms is used at 2.4 tonnes per hectare. The total grain yield per year (long rain plus short rain) is 5 tonnes per hectare. The baseline scenario assumes a 29 percent maize yield increase on the plot area receiving the 18 tonnes of carbon per hectare biochar application. For a 5 tonnes per hectare yield for the control, a 29 percent increase results in a yield rate of 6.5 tonnes per hectare with biochar additions. The actual measured yield increases varied significantly with changes in soil fertility (Kimetu et al. 2008) and ranged from –18 percent to +97 percent compared to the control for the same period. The sensitivity analysis considers a range in crop yield effect from –50 percent to +97 percent, which allows for exploring the variability in the extent of the preexisting soil degradation, time since forest conversion, farmer practices, weather conditions, and fertilization rates. The maximum value of +97 percent increase is based on the Kimetu et al. (2008) study for full nitrogen, phosphorus, and potassium fertilization on control plots in highly degraded sites (80–105 years after forest conversion). Reviews of biochar applications rates and plant responses have shown a wide variation, from –29 percent to +324 percent (Glaser, Lehmann, and Zech 2002), or –60 percent to +100 percent (Verheijen et al. 2010).

The increased grain yields would result in an increase in food available for the family. In the case of subsistence farming (as applies for farmers in this region), any surplus grain is used to feed the family. It is unusual for a surplus to be sold at markets, as the primary concern is providing enough food for self-sufficiency of the household. In the rare case where there is excess beyond what the family needs, farmers may not be able to sell at market prices due to the role played by intermediaries.[7] Therefore the baseline scenario assumes no revenues from increased maize yields. However, the economic analysis and discussion also considers a scenario where this surplus in maize is monetized using the August 2010 price, whereby farmers receive 2,000 Kenyan shillings (about $24) for one 90-kilogram sack of air-dried, shelled maize grain. The sensitivity analysis considers a range in maize prices, as the price of maize grain is extremely fluid depending on the time of year (increased price during the dry season) and the general success of the harvest.[8]

Duration of Biochar's Agronomic Effect

For the soils in this region of Kenya, biochar's effect on crop productivity was found to be due mainly to increased soil organic carbon (Kimetu et al. 2008), and thus would be expected to last for the lifetime of the biochar in the soil. The baseline scenario assumes a 50-year agronomic effect for the applied biochar, which is a very conservative assumption based on the expected mean residence times of biochar as cited above. The sensitivity analysis investigates this aspect.

Soil Organic Carbon from Residue Removal

When crop residues such as maize stover are removed from the field, there is the potential to decrease soil organic carbon as compared to fields without stover removal (Moebius-Clune et al. 2008). Kapkiyai et al. (1999) in Kenya report that for plots receiving no other inputs, after 18 years the soil organic carbon is 23.8 tonnes of carbon per hectare with stover removal and 24.6 tonnes with stover remaining. The difference is 0.8 tonnes of carbon per hectare loss over 18 years, or 0.04 tonnes of carbon annual average for complete stover removal. Thus, with 42 percent of total residue collection for pyrolysis, 0.02 tonnes of carbon per year is lost. This corresponds to 4 kilograms of CO_2e per tonne of dry stover removed, based on an average stover yield of 17.4 tonnes per hectare for the long and short rain seasons across all sites. The issue of erosion protection with crop residue removal is discussed under the "Feedstock" subsection above.

Soil Organic Carbon from Increased Productivity

As crop productivity increases with biochar application, the amount of aboveground and belowground biomass increases. Because aboveground biomass may be collected for pyrolysis or other on-farm needs, the LCA model assumes only the increase in belowground biomass contributes to an increase in soil organic carbon levels. Shoot to root ratio measurements of stover biochar-amended maize plants are around 0.9 (Torres et al. 2011). The amount of soil organic carbon from the incremental belowground biomass is estimated to be 0.04

tonnes of carbon per hectare per year, as discussed in the preceding paragraph. The average long rain stover yield (2005–09) of 4.6 tonnes per hectare and the 0.9 shoot to root ratio is used to calculate a soil organic carbon accumulation of 0.008 tonnes of carbon per tonne of belowground biomass. For a 29 percent crop yield increase and an average stover to grain ratio of 2.7 across all sites for 2005–09,[9] the increase in stover per year is 2.1 tonnes per hectare, and thus the increase in belowground biomass per year is 2.3 tonnes per hectare. At this level, the incremental soil organic carbon increase due to increased belowground biomass productivity with biochar additions is 0.02 tonnes of carbon per hectare per year. This corresponds to 2.7 kilograms of CO_2e per tonne of biochar per year, or 134 kilograms of CO_2e per tonne of biochar for a 50-year biochar effect.

Soil Organic Carbon from Priming Effects
The LCA does not include positive or negative priming effects. This is due to the fact that the controlling mechanisms are not yet fully known, and both short-term positive and long-term negative priming have been found. More detailed discussion on priming effects can be found under "Soil Organic Matter" in the section titled "Impacts on Soil Health and Agricultural Productivity" in chapter 3.2.

The Role of Nitrogen and Biochar
Nitrogen plays a critical role in agriculture and the role of biochar application and its effect on nitrogen must be discussed. The changes in nitrogen dynamics relevant to crop productivity are included implicitly as the crop yields are an aggregate result of the intervention. Three aspects are of concern: (a) the nitrogen changes in the biochar; (b) the nitrogen emissions during pyrolysis; and (c) the nitrogen changes in the soil after biochar application.

When crop residues are removed from the field, the nitrogen they contain is also removed. Research in the Lehmann group at Cornell continues to investigate the nitrogen dynamics in these systems in Kenya (Torres et al. 2011). In this study, the fresh stover biomass was found to have a nitrogen content of 0.58 percent. For the annual supply of secondary feedstock (1.87 tonnes) for the household pyrolysis cookstove, about 11 kilograms of nitrogen are removed with the residue. During pyrolysis, approximately half of the nitrogen is retained in the biochar, and the relative nitrogen content of the biochar actually increases to 1.22 percent while the C:N ratio decreases from 80 to 54 (Torres 2011). Thus, when the 0.52 tonnes of biochar are applied back to the soil annually, 6 kilograms of nitrogen are returned with it.

Second, the nitrogen species emitted during pyrolysis are assumed to be in the form of nitrogen oxides, as nitrous oxide is typically associated with high-temperature combustion, and has been demonstrated to be a small fraction of biomass stove GHG emissions in other studies (Johnson et al. 2008; MacCarty et al. 2008; Zhang et al. 2000). The nitrogen oxide emissions data for this stove are calculated to be 0.0036 kilograms of nitrogen oxides per household per year, or 0.00065 grams per kilogram of fuel. This is significantly lower than the range of 0.4 to 1.1 grams per kilogram of fuel found for a series of cookstoves in China (Wang 2007). Although nitrogen oxides are presently assumed to be greenhouse neutral, it is a precursor to ozone and can result in acid rain (MacCarty et al. 2008).

Because of the high C:N ratio of biochar made from wood as in this project (C:N of 387), another important consideration is the potential for nitrogen immobilization with biochar application. The process of pyrolysis forms hetero-cyclic nitrogen compounds, meaning that the nitrogen is largely unavailable despite the decrease in C:N ratio with pyrolysis. In addition, soil microbial communities require nitrogen in order to utilize carbon. If there is not sufficient nitrogen in the biomass (or biochar), then the microbes take nitrogen from the soil, making it unavailable to the plant. The nitrogen immobilization with biochar application would be expected to be a short-term or immediate effect that would potentially decrease crop yields. However, it is likely that the crop yield data incorporate this effect, thus any reduced nitrogen availability is assumed to be taken into account. (Torres [2011] actually found that there was an increase in crop uptake of nitrogen with biochar addition compared to fresh biomass addition. It is hypothesized that the increased nitrogen uptake signaled a lower nitrogen immobilization, not due to improved nitrogen availability, but rather due to reduced carbon availability because of the highly recalcitrant nature of carbon in the biochar.)

Another consideration is the role of biochar on fertilizer use efficiency. Studies have suggested an increase in nitrogen fertilizer use efficiency with biochar additions (Chan et al. 2007; Steiner et al. 2008; van Zwieten, Kimber, Downie et al. 2010). This effect would potentially increase crop yields and is assumed to be incorporated into the effect of biochar on crop yields. The effect could decrease the amount of nitrogen fertilizer applied to soils. However, because fertilizer inputs on these farms are generally low, it is likely that the amount applied would not change but rather their effectiveness would be improved. The nitrogen immobilization and nitrogen fertilizer use efficiency may potentially be counteracting the effects of biochar on crop production. Although crop yield data are available only in the shorter term (five years), it is possible that in the longer term the effects of one effect may become more pronounced. As longer-term field trials continue, these interactions will become better understood.

Soil Nitrous Oxide and Methane Emissions

The baseline scenario of this analysis assumes that there are no changes in soil nitrous oxide emissions with biochar application for the following reasons: (a) lack of site-specific data for fertilizer-related nitrous oxide emissions; and (b) variability in fertilizer application rates on smallholder farms in Kenya. However, biochar reportedly reduces nitrous oxide soil emissions in many situations that result from nitrogen fertilizer application (see the section titled "Impacts on Climate Change" in chapter 3). The sensitivity analysis considers a range of +50 percent to –50 percent for the effect of biochar application on soil nitrous oxide emissions. The current estimate for the conversion of fertilizer nitrogen to nitrous oxide emissions of 1.325 percent for U.S. corn production (Wang 2007) is used in the sensitivity analysis.

This analysis does not assume any effect of biochar on methane emissions from soil. Experiments have indicated that not all biochars will affect nitrous oxide or methane emissions equally (see the section titled "Impacts on Climate Change" in chapter 3 for further detail).

Avoided Nitrous Oxide and Methane Emissions from Biomass
This analysis assumes no additional emissions of non-carbon dioxide GHGs as a result of biomass additions to soil.

Results and Discussion

Table 5.3 lists the results for the Kenya pyrolysis cookstove system. The net GHG balance is −1.8 tonnes of CO_2e per tonne of dry matter, as illustrated further in figure 5.3. The amount of time saved with the pyrolysis cookstove is 62 hours per tonne of dry matter, and the net economic balance is −$1.43 per tonne of dry matter. The majority of both the GHG emissions and reductions are from carbon dioxide, with much smaller contributions from methane and nitrous oxide (table 5.3). The surplus maize from biochar additions is 0.74 tonnes of grain per tonne of dry matter over the 50 years of biochar's effect on crop yields.

Looking at the contribution analysis in figure 5.3a, the largest source of reduced GHG emissions is the avoided traditional cooking, at −1.3 tonnes of CO_2e per tonne of dry matter, or 69 percent of the total GHG reductions. This large contribution is due mainly to the fact that 49 percent of the traditional cooking fuelwood is collected off-farm, of which 80 percent is from a nonrenewable source. Meanwhile, all of the primary and secondary feedstock for the pyrolysis cookstove is able to be supplied on-farm, and is thus considered renewable.

Table 5.3 Result Vector for Pyrolysis Cookstove in Kenya Baseline Scenario per Tonne of Dry Feedstock

Input/output	Value	Unit
Off-farm wood fuel	0.93	tonne
Time	61.5	hour
Methane	− 5.55	kg
Nitrous oxide	7.82E-07	kg
Carbon dioxide	−1,628	kg
Stable carbon	147	kg
Net CO_2e	**−1,767**	**kg**
Revenue	0	$
Cost	−1.43	$
Net $	**−1.43**	**$**
Surplus maize	0.74	tonne

Source: World Bank.
Note: For methane, nitrous oxide, and carbon dioxide, a negative sign corresponds to avoided emissions or sequestration, and a positive value indicates emissions. Negative $ are costs, and positive $ are net revenues. Positive time corresponds to time savings. The bold type indicates the net GHG and economic balances.

Figure 5.3 Sources of GHG Reductions and Emissions, and Carbon Balance for Secondary Feedstock

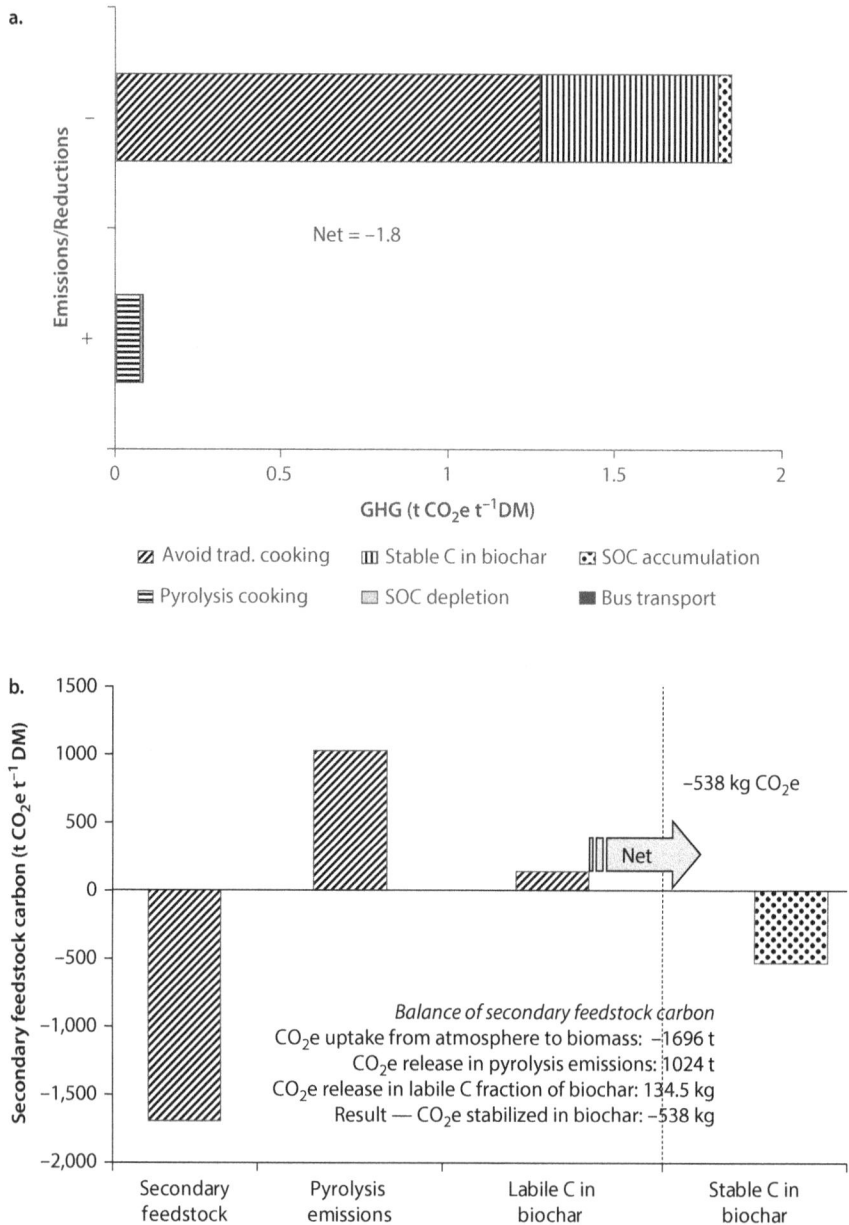

Net = −1.8

GHG (t CO_2e t^{-1}DM)

Emissions/Reductions

☒ Avoid trad. cooking   ▥ Stable C in biochar   ⬚ SOC accumulation

▤ Pyrolysis cooking   ▥ SOC depletion   ■ Bus transport

Secondary feedstock carbon (t CO_2e t^{-1} DM)

−538 kg CO_2e

Net

Balance of secondary feedstock carbon
CO_2e uptake from atmosphere to biomass: −1696 t
CO_2e release in pyrolysis emissions: 1024 t
CO_2e release in labile C fraction of biochar: 134.5 kg
Result — CO_2e stabilized in biochar: −538 kg

Secondary feedstock   Pyrolysis emissions   Labile C in biochar   Stable C in biochar

Source: World Bank.

Note: 5.3a. Figure shows contribution analysis for the net climate change impact per tonne of dry secondary feedstock for a pyrolysis cookstove biochar system in Kenya. The upper bar (−) represents the GHG reductions, the lower bar (+) is GHG emissions, and the difference represents the net GHG balance of the system. C = carbon; SOC = soil organic compound; CO_2e = carbon dioxide equivalent. 5.3b. Figure shows carbon balance for secondary feedstock. The pyrolysis carbon dioxide emissions associated with the secondary feedstock and the labile carbon in the biochar are subtracted from the quantity of CO_2e taken up by the renewable, secondary feedstock. The net result is the CO_2e from the stable carbon in the biochar. This schematic demonstrates how the pyrolysis CO_2e emissions from the secondary feedstock are accounted for. Thus, in 5.3a, the pyrolysis emissions are the non-carbon dioxide GHG emissions only. C = carbon; CO_2e = carbon dioxide equivalent.

Therefore, the carbon dioxide emissions from the pyrolysis cookstove are cancelled out by the uptake of carbon dioxide of the renewable feedstocks, highlighting the important role that the renewability of the feedstock plays in the GHG balance. Only non-carbon dioxide GHG emissions are included in the pyrolysis emissions, as these are assumed to not otherwise occur. The carbon balance for the secondary feedstock (crop residues) is illustrated in figure 5.3b. The stable carbon in the biochar is responsible for the sequestration of an additional −0.5 tonnes of CO_2e per tonne of dry matter (29 percent of total reductions).

The greatest amount of GHG emissions occurs during pyrolysis cooking (+0.8 tonnes of CO_2e per tonne of dry matter, or 95 percent of total emissions). The 95 percent of emissions seems high, but it is actually quite small, as the total emissions are negligible in comparison to the total reductions. The soil organic carbon accumulation (due to increased crop productivity) and soil organic carbon depletion (due to crop residue removal) contribute only minor amounts to the balance, at 2 percent of the GHG reductions and 5 percent of the GHG emissions, respectively. Finally, transportation has the smallest impact, at less than 0.01 percent.

From the GHG contribution analysis it is clear that the avoided traditional cooking is the primary source of net GHG reductions. However, as mentioned above, the sustainable harvest of both primary and secondary feedstocks are critical to ensure carbon dioxide uptake with regrowth. To compare the GHG balance of −1.8 tonnes of CO_2e per tonne of dry matter to other studies, a different functional unit is used in the following, which is the amount of cooking energy required for one family for a year. This results in the following GHG balance: −3.3 tonnes of CO_2e per household per year.[10] It is possible that comparable GHG reductions could be achieved with nonpyrolytic improved cookstoves. For example, 2.3–3.9 tonnes of CO_2e per household per year of emission reductions could be achieved for improved cookstoves in Mexico (Johnson et al. 2009). However, within the emission reductions of the pyrolysis cookstove system, atmospheric carbon dioxide is converted to the stable carbon in the biochar. On the other hand, energy efficiency of the pyrolysis stove is lower (Whitman et al. 2011). The pyrolysis cookstove has the benefit of utilizing secondary fuel, which reduces fuelwood consumption, fuelwood collection time, and deforestation pressures even further than an improved cookstove that still requires only primary feedstock. In addition to the monetary inputs and outputs, the LCA also calculates the net time savings, which is 115 hours per household per year, or 65 hours per tonne of dry matter. The net time savings estimate takes into account the reduced fuel collection due to decreased primary fuel consumption and the increased time to apply biochar to the fields. The time saved could be utilized for other family, farming, or community activities. In 2008, the lower agricultural minimum wage was 2,536 Kenya shillings ($35) per month for Kenya (U.S. Department of State 2009). Assuming a workweek of 52 hours, this translates to $0.16 per hour, which corresponds well with the $1.25 per day rate for agricultural workers in Kenya.[11] The 115 hours saved per year would then correspond to an additional $18.40 per year for a net revenue of +$16 per household, or $8 per tonne of dry matter, compared to the baseline −$1 per tonne of dry matter without valuing the time saved (figure 5.4).

Figure 5.4   Contribution Analysis for Net Revenues per Tonne of Dry Secondary Feedstock for Pyrolysis Cookstove Biochar System in Kenya

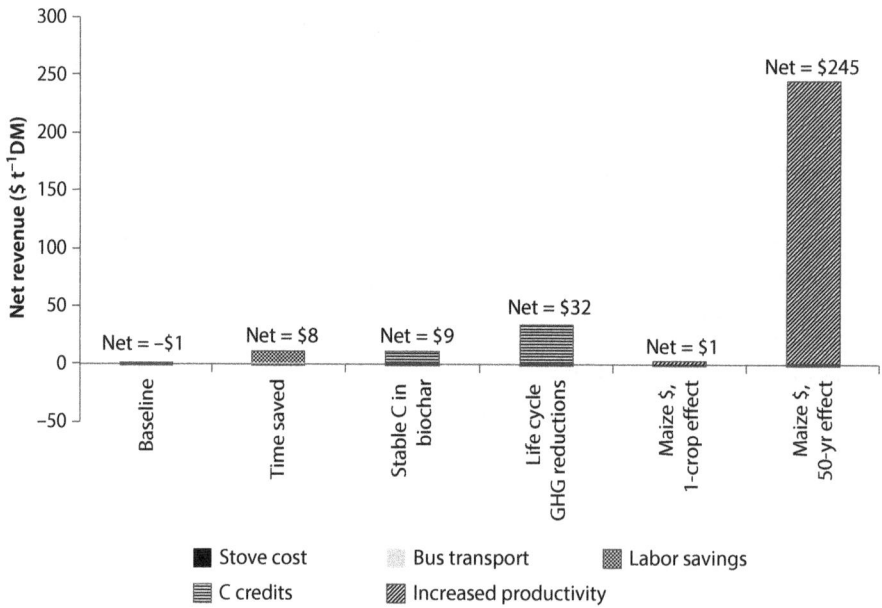

Source: World Bank.
Note: The six scenarios are as follows: "baseline" includes only the monetary inputs and outputs of the baseline scenario; "time saved" is the baseline plus a monetary value assigned to the time saved in fuelwood collection; "stable carbon in biochar" values GHG offsets for the carbon dioxide sequestered directly by stable carbon in biochar; "life-cycle GHG reductions" values the net GHG reductions, which includes avoided emissions; "maize $, 1-crop effect" is where the surplus maize produced is monetized for a one-crop biochar effect; and "maize $, 50-yr effect" is where the surplus maize produced is monetized for a 50-year biochar effect.

The pyrolysis cookstove system also offers the advantage of improved soil fertility over nonbiochar cookstove scenarios. The baseline scenario does not capture the monetary value of this improved fertility because of the reality of subsistence farming—surplus grain feeds the family but does not usually generate any income. However, the two scenarios, "maize $, 1-crop effect" and "maize $, 50-yr effect" of figure 5.4 explore this, where a monetary value is assigned to the surplus maize. The baseline LCA assumes the 50-year biochar effect, which would translate to a gain of $246 per tonne of dry matter from surplus maize sales, for a net of $245 over the 50 years of biochar's effect. Assigning a maize value for a one-crop biochar effect results in a net of $1.

A biochar payback period provides an estimate of the number of years it would take biochar used as a soil amendment to pay for itself. If the surplus maize, labor savings, or carbon offsets are not monetized, then the payback will never take place in monetary terms (that is, the net revenue will always be negative). However, this would not be an accurate assessment of the improved livelihoods of these smallholder farmers. For this reason, the payback period is calculated assuming the base price of $24 per 90-kilogram bag of maize grain. When monetizing the surplus maize, the biochar would pay for itself in the first year.

Finally, the possibility of pyrolysis cookstove programs receiving carbon credits would further improve the livelihoods of smallholder farms, as demonstrated in figure 5.4. For a carbon dioxide price of $14,[12] or $19, assuming an exchange rate of $1.36 to the euro, the revenues could be an additional $33 per tonne of dry matter (for a net of $32) when valuing the net GHG reductions, or an additional $10 (for a total of $9) if only the stable carbon in the biochar were eligible for carbon credits.

The 3.3 tonnes of CO_2e per household per year net GHG reductions is within the range of values calculated in the Stove Impact on Climate Change Tool (SImpaCCT) systems dynamics model from Whitman et al. 2011, where between 2.6 and 5.8 tonnes of CO_2e per household per year of GHG reductions were modeled. The lower end of this range (2.6–4.7 tonnes) from SImpaCCT is based on a prototype pyrolysis stove, and the upper range (3.3–5.8 tonnes) is for a refined pyrolysis cookstove. The data for the present LCA utilizes the prototype cookstove data, which sit centered in this lower range. The 4.7 tonnes of CO_2e per household per year calculated in SImpaCCT assumes that 50 percent of on-farm available residues are collected and all off-farm biomass collected is nonrenewable. However, one of the primary differences is that SImpaCCT utilizes a higher fuel consumption rate than the present LCA (discussed above in the "Pyrolysis and Cooking" subsection). With the lower fuel consumption rate used in the LCA, all of the biomass needs are met on the farm, thus no off-farm fuel is collected for pyrolysis cooking in the baseline scenario (but avoided traditional cooking requires 49 percent of fuel needs to be met with off-farm fuelwood, of which 80 percent is nonrenewable). In addition, although the present LCA assumes biochar is agronomically effective for 50 years (which contributes –0.04 tonnes of CO_2e from increased soil organic carbon), LCA is a static modeling tool that does not incorporate temporal dynamics and feedback mechanisms. Meanwhile, SImpaCCT is a dynamic model that considers changes in stocks and flows of carbon over time.

Alternative Functional Units
Another viewpoint for thinking about the long-term effect of biochar on crop productivity would use 1 tonne of biochar as the functional unit. This allows the costs and benefits to be assigned on a per biochar basis and still accumulate long-term effects for the duration of biochar's effectiveness, without using a functional unit with a temporal basis that may potentially confuse the issue. With this in mind and utilizing a 50-year effect of biochar on soil properties, the net impact of 1 tonne of biochar for the Kenya cookstove system results in GHG reductions of –6.3 tonnes of CO_2e and 2.7 tonnes of surplus maize per tonne of biochar. If the surplus maize is not monetized, the net economic balance is –$5, whereas if the surplus maize is valued at $24 per bag, then the balance is $879 per tonne of biochar. Again, these results are for a 50-year biochar effect on crop productivity. It is not assumed within the lifetime of this project that more biochar is applied than the specified application rate, or that crop yields decrease with increasing biochar application rates, as the maximum tested is not exceeded.

Table 5.4 Sensitivity Analysis Input Parameters, Including Baseline and Range Values

Parameter	Baseline	Sensitivity range
Primary feedstock (tonnes of dry matter per household per year)	1.15	1.15–3.02
Stable carbon content of biochar (%)	80	0–90
Yield response with biochar additions (%)	+29	−50 to +97
Maize prices ($ per 90-kg bag)	0	18–36
Duration of biochar's effect (years)	50	1–100
Pyrolysis methane emissions (kg per tonne of dry matter)	1.98	1.64–6.4
Three-stone fire methane emissions (kg per tonne of dry matter)	5.17	0.60–6.4
Fraction of nonrenewable biomass for off-farm wood	0.8	0–1.0
Biochar application rate (tonnes per hectare)	27	2.7–27
Soil nitrous oxide emissions (%)	0	−50% to +50%
Fraction of secondary feedstock collected off-farm	0	0–1.0

Source: World Bank.

Sensitivity Analysis

A sensitivity analysis was conducted in order to measure the variability in the LCA results as a function of varying key input parameters (table 5.4). The input parameters that were tested were the primary feedstock consumption, the fraction of recalcitrant carbon in the biochar, the yield response of maize crops with biochar additions, the price the farmer receives for maize, methane emissions of the pyrolysis cookstove, methane emissions of the traditional three-stone cookstove, the duration of biochar's agronomic effectiveness, soil organic carbon accumulation, the fraction of nonrenewable biomass for off-farm wood, the biochar application rate, soil nitrous oxide emissions, and the fraction of secondary feedstock collected off-farm. The detailed sensitivity

Figure 5.5 Sensitivity Analysis for Net GHG per Tonne of Dry Secondary Feedstock for Pyrolysis Cookstove Biochar System in Kenya

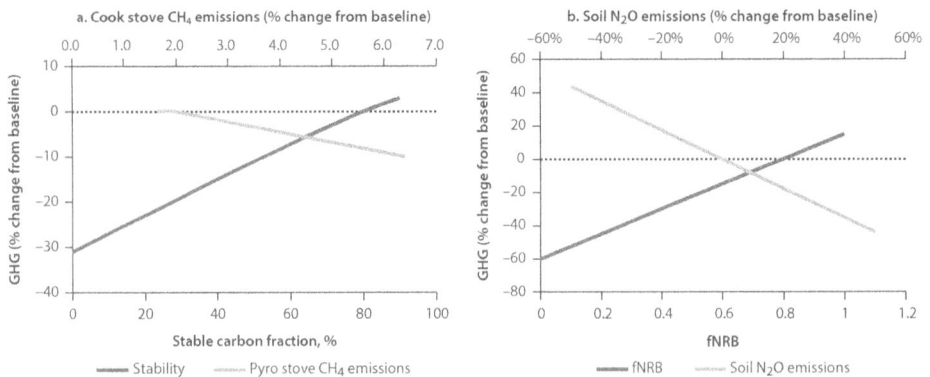

Source: World Bank.
Note: Figure 5.5a compares the fraction of stable carbon in the biochar to the pyrolysis cookstove methane emissions, and 5.5b illustrates the variation in fraction of nonrenewable biomass and soil nitrous oxide emissions (nitrous oxide emissions range chosen according to extreme values available in the literature, see text). CH_4 = methane; N_2O = nitrous oxide; GHG = greenhouse gas; fNRB = fraction of nonrenewable biomass.

Figure 5.6 Sensitivity Analysis for Net Economic Balance per Tonne of Dry Secondary Feedstock for Pyrolysis Cookstove Biochar System in Kenya

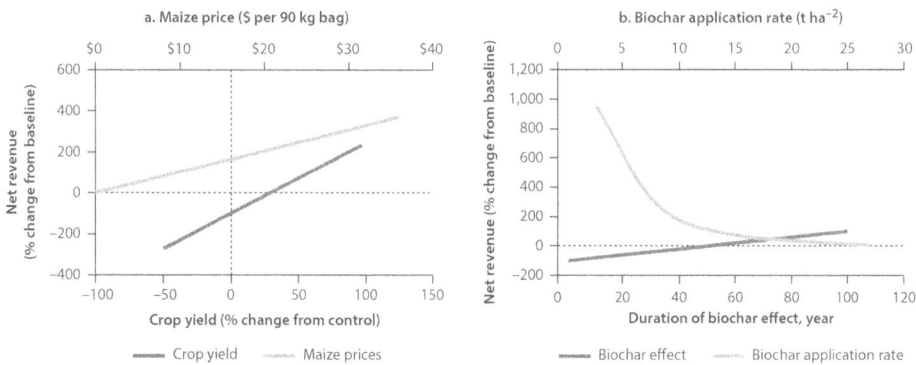

a. Maize price ($ per 90 kg bag)

b. Biochar application rate (t ha^{-2})

Source: World Bank.

Source: World Bank.

Note: Figure 5.6a compares the crop yield to maize prices, and 5.6b illustrates the variation in the duration of biochar's agronomic effect and the biochar application rate.

analysis is presented in appendix B, while a summary of the sensitivity analysis results is described here.

The GHG balance is relatively insensitive (less than around 10 percent variability) to the pyrolysis methane emissions, the avoided traditional cooking methane emissions, the duration of biochar's agronomic effect, and the crop yield response to biochar additions. The fraction of the stable carbon in the biochar within the realistic range of 50–90 percent influences the net GHG balance by only 11 percent, and thus also has a relatively small influence on the net results. Meanwhile, the GHG balance is more sensitive to the primary feedstock consumption rate (100 percent), the fraction of nonrenewable biomass for off-farm wood (60 percent), the biochar application rate (20 percent), and soil nitrous oxide emissions (±44 percent) (figure 5.5).

Significant effects on the net economic balance are found within the range tested for the duration of biochar's agronomic effect (100 percent), crop yield response (274 percent), biochar application rate (956 percent), and how the surplus maize crop is monetized (369 percent). Figure 5.6 illustrates this variability, where it is evident that the biochar application rate has the largest impact on the net economics, followed by the crop yield response due to biochar additions. The duration of biochar's effect appears small next to the biochar application rate, but still results in up to 100 percent change in the net economics, while the variability in maize prices affects the results by up to 369 percent. More details on the sensitivity analysis can be found in appendix B.

Additional Considerations

There are specific issues unique to biochar systems that are not incorporated into the LCA because they are outside the system boundaries or scope of the analysis or not quantifiable in a traditional sense. However, they are important to include as a discussion item. One challenge is in handling biochar. The high volume to weight ratio means there is a potential for biochar to be blown or washed away, and the application to soils can be difficult. Inhalation of fine airborne biochar particles is also an important concern. If not quenched or cooled sufficiently, biochar is highly flammable and could present a fire hazard (section titled "Impacts on Climate Change" in chapter 3). The families operating the stoves would need to be aware of these issues to ensure that the biochar is completely cooled or quenched prior to removing from the cookstoves. Another concern is storing the biochar. Because small quantities are produced at a time, the family may choose to store the biochar until sufficient amounts are collected and it is the best time for application. However, if the biochar is applied right away, the need to store the biochar would be eliminated. Another valid concern, as discussed by Iliffe (2009), is that the quantity of biochar produced is limited by the amount of cooking. Even if the family wanted to produce more biochar, the cookstove remains hot for a few hours after use and cannot be handled to remove the biochar until it is cooled. Thus, the number of meals cooked per day limits the amount of biochar available to the family. However, due to renewable feedstock limitations, this may be the most sustainable scenario until other feedstock sources are identified.

Box 5.2 Summary of Kenya Case Study

The Kenya household pyrolysis cookstove system with biochar returned to soil has the potential for climate change mitigation through carbon sequestration and GHG emission reductions, while also being economically viable for the smallholder farmer. Using the pyrolysis cookstove for household cooking and producing biochar, the net GHG reductions are –1.8 tonnes CO_2e per tonne of dry pyrolysis (secondary) feedstock. The net economic balance is –$1 per tonne of dry matter where no monetary value is assigned to surplus grain, labor savings, or potential carbon credits. If the 62 hours saved in fuelwood collection are monetized, the net return would be $8 per year. If carbon credits were received (at $19 per tonne of CO_2e), the household could return from $27 to $50 per tonne of dry matter, depending on whether avoided emissions are valued the same as carbon sequestered directly in the biochar. If the surplus maize is valued, the net return would be $1 per cropping season of biochar's effectiveness, or $245 for a biochar effectiveness of 50 years. Synergistic benefits of the pyrolysis cookstove project may include decreased indoor air pollution and related illnesses (section titled "Social Impacts" in chapter 3), reduced fuelwood consumption and thereby decreased deforestation pressures (section titled "Impacts on Climate Change" in chapter 3), and improved long-term soil fertility (section titled "Impacts on Soil Health and Agricultural Productivity" in chapter 3).

The contribution analysis reveals that the net GHG balance is largely driven by the avoided emissions from the traditional three-stone fire and the sustainability of the primary and secondary cookstove feedstock. If the feedstock is unsustainably harvested and does not regrow, then emissions during cooking would not be offset by the biomass regrowth. Meanwhile, the amount of stable carbon in the biochar contributes 29 percent of the net GHG reductions. Soil nitrous oxide emissions can also play an important role in the GHG balance, thus more data are required to quantify this effect.

The sensitivity analysis reveals that the net GHG balance is relatively insensitive to changes in methane emissions during traditional or pyrolysis cooking, the crop yield response with

box continues next page

Box 5.2 Summary of Kenya Case Study *(continued)*

biochar additions, and the duration of biochar's agronomic effect, within the realistic range tested for these parameters. The surplus maize is most sensitive to the crop yield response and the biochar application rate. Meanwhile, the net revenues are very sensitive to the crop yield response and the price of maize.

This Kenya pyrolysis cookstove system presents itself as a low-risk biochar project with climate change adaptation and economic benefits for the smallholder farmers implementing these stoves and utilizing the biochar.

Vietnam Case Study Life-Cycle Assessment

System Overview

The biochar project described here is a component of a larger development project related to institutional capacity building, irrigation water management, and soil nutrient management. The title of the project is "Improving the utilization of soil and water resources for tree crop production in coastal areas of Vietnam and New South Wales," and it is funded by the Australian Centre for International Agricultural Research and implemented by the Agricultural Science Institute for Southern Central Coastal Vietnam in collaboration with the Department of Primary Industries, New South Wales, Australia. The project started in 2007 and aims to improve incomes of low-income farmers in central coastal Vietnam, where large areas of sandy soils occur. Charcoal, or biochar, is produced as a by-product of spring roll rice wafer cottage industries in the Ninh Thuan province in the central coast region. The feedstock for the rice wafer stoves is rice husks, which would otherwise be burned as a means of waste management. The biochar by-product is purchased from local families by the project working with three different farmers in the area, and is applied to peanut and cashew fields. The case study uses data from peanut farms, as the project contacts indicate that there is a trend of farmers transitioning to growing more peanuts than cashews because of higher potential revenues, and the peanut data are more mature than the cashew data at the time of writing. The biochar is applied wet, at 20 tonnes per hectare, or 12 tonnes of dry biochar per hectare.

Methodology

Function, Functional Unit, and Reference Flows

For the Vietnam case study, biochar production is a multioutput system with up to four coproducts: biomass management, soil amendment, carbon sequestration, and bioenergy production. The functional unit of the system is 1 tonne of dry rice husk, which is used as a rice wafer stove feedstock. The subsequent biochar is applied to peanut fields on a farm in Vietnam.

System Boundaries

A flow diagram of the biochar system is illustrated in figure 5.7. The system is organized into four modules: rice husk feedstock, cooking, biochar, and crop response, each of which has multiple subprocesses.

Figure 5.7 Schematic Flow Diagram for Vietnam Biochar System

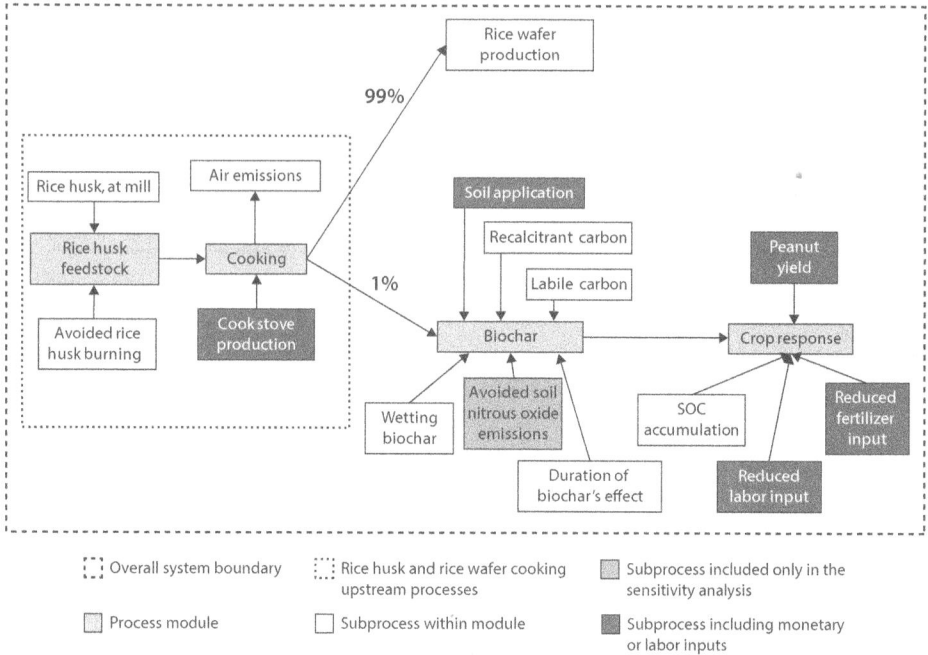

Source: World Bank.
Note: SOC = soil organic compound.

Because in this case study the biochar production is a by-product of an established rice wafer cooking business, economic allocation is used to separate the impacts associated with cooking and the rice husk feedstock of the biochar production from the rice wafer production. The allocation is done by taking the relative percentage of biochar to the total products (biochar + rice wafers) on an economic basis, then including that percentage impact for the upstream processes, as illustrated in the process flow diagram. This affects both "positive" (avoided emissions from rice husk burning) and "negative" (emissions from cooking) upstream impacts equally, and the relative allocation is tested via the sensitivity analysis. All other processes in this LCA are included via the system expansion method.

Life-Cycle Inventory
Rice Husk Feedstock
The project contacts state that local reports indicate about 50 percent of rice hulls at mills are used for a range of downstream purposes (such as for fuel in rice mills, in homes, and in brick kilns), and about 50 percent represent a waste problem (and are thus burnt for disposal) (Slavich et al. 2010). The rice husks are typically piled up at rice mills and eventually burned if not used for other purposes. Hence, utilizing the rice husk as a fuel is also a waste management strategy. Thus, this LCA assumes that if the rice wafer stoves were not using the rice husks, then the rice husks would be burned, as in the "avoided rice husk burning" subprocess.

The rice husk feedstock is available in 200-liter bags. Using an average rice husk density of 90 kilograms per cubic meter,[13] each bag contains 18 kilograms of rice husk. One 200-liter bag of rice husk costs 6,000 Vietnamese dong (VND) ($0.27), which includes the cost of the bag (3,000 VND) and filling the bag. If farmers were to bring their own bag and fill it themselves, the cost would be reduced to only 500 VND per bag.

Avoided Rice Husk Burning

Rice husks not utilized are piled up at rice mills and ultimately burned in piles as a waste management strategy. Emission factors were taken from data for smoldering rice husk mounds used in charcoal production in Thailand. These mounds have layers of rice husks covered by a layer of soil over a charge of approximately 200 kilograms of wood (Smith et al. 1999). The rice husk piles are the most representative data of in-field rice husk burning available, and these emission factors will be explored in the sensitivity analysis. The carbon dioxide, carbon monoxide, methane, and nitrous oxide emission factors from the Smith et al. study are 443, 30.9, 3.71, and 0.0232 kilograms per tonne of dry feedstock, respectively. In addition to the GHG emissions from residue burning, a value of 3 percent of the carbon in the residue is estimated to be converted to char during open residue burning (Forbes, Raison, and Skjemstad 2006; Glaser, Lehmann, and Zech 2002).

Rice Wafer Stove Construction

The rice wafer stove is a slanting grate updraft gasifier stove with a chimney. There are few data on the construction of the rice wafer stove beyond what is evident from qualitative descriptions and photographs (figure 5.8). These are

Figure 5.8 Rice Wafer Stove in Vietnam

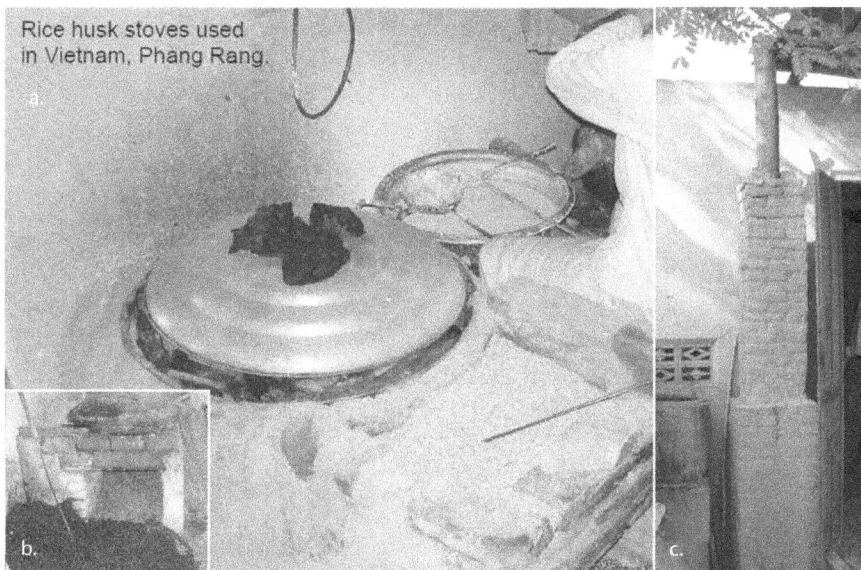

Source: P. Slavich (2010).
Note: Figure 5.3a shows a top view of the rice husk feedstock input compartment and rice wafer cooking area; 5.3b is a side view where biochar is raked out; 5.3c is an outdoor view of smokeless chimney.

traditional stoves that have been in use for at least 100 years. There are more than 100 households with similar stoves in the village, and this would be the same for villages throughout the region.[14] The stove is indoors and located against an exterior wall. The indoor section consists of the cooking and feedstock input areas (figure 5.8a), and the chimney is outdoors (figure 5.8c). The materials for constructing the stove are estimated from the number of bricks, where approximately 622 kilograms of brick are required per stove. The amount of mortar is estimated from the approximate mortar required of 1,300 kilograms per 1,000 bricks with ½-inch joint spacing,[15] which comes to 810 kilograms of mortar per stove. The mortar is assumed to be a clay-based material. The stove lifetime is estimated at 20 years for the baseline scenario.

With these parameters and the economic allocation (discussed below), the input materials required per functional unit are only 5 kilograms of brick and 7 kilograms of mortar. These are below the cutoff requirements of input materials (5 percent of the mass of the functional unit), thus for this reason it is conservative to assume that the processes further upstream can be excluded from the LCA (that is, brick or mortar production or transportation). A baseline of $100 is used for the cost of the stove. The equivalent annual capital cost of the stove is calculated based on the 20-year lifetime and a 10 percent discount rate. The stove is built on-site, and thus transportation of the stove is not required.

Cooking: Biochar and Rice Wafer Production
The rice husk feedstock is purchased by the rice wafer-producing family in 200-liter mesh bags and is transported approximately 5 kilometers via bullock cart from the rice mill to the home. The rice wafer stove is operated continuously over a 10–12-hour period. The rice husk feedstock is poured into a hopper, and the biochar is manually raked out approximately every hour. The biochar is wet by the stove user to prevent autoignition and then bagged. After the biochar is wet, it is put into the same bags it came in. One 200-liter bag of rice husk lasts about three hours, thus approximately four bags of rice husk are consumed per day, or 26 tonnes per year. About one 200-liter bag of biochar is produced from five to six bags of rice husks. The biochar yield is 35 percent by dry weight.[16]

Rice husk biochar has very low density, at 0.15 grams per cubic centimeter (Haefele et al. 2009). The moisture content of the wet biochar is approximately 40 percent on a wet basis,[17] and 0.7 liters of water is used to wet 1 kilogram of dry biochar. Thus, a 200-liter bag of wet biochar weighs about 50 kilograms. The rice wafer-producing family sells the bag of biochar for 4,000 VND ($0.18). This corresponds to $3.64 per wet tonne of biochar, or $6.06 per dry tonne of biochar.

Emissions Data
Emissions measurements for the rice wafer stove have not been conducted at this stage. Project contacts and photographs indicate that the stove is flued outside the building and exhibits smokeless operation. Because of the lack of stove-specific data, emissions are calculated based on experimental data from a Mayon

Figure 5.9 Mayon Turbo Rice Hull Stove

Source: MacCarty et al. 2007.

Turbo rice hull stove (MacCarty et al. 2007), as in figure 5.9. This stove does not have a flue, and is of quite different construction from the Vietnamese rice wafer stove. MacCarty et al. (2007) compared six stoves, and found a correlation in emission ratios and species with feedstock type. For this reason rice hull-specific data are used. The stove emissions are assessed in the sensitivity analysis.

The emissions calculations utilize the MacCarty et al. 2007 emission ratios of carbon dioxide and products of incomplete combustion, similar to the methodology in Whitman et al. 2011 and for the Kenya cookstove case study. Because the biochar and cooking energy yields of the rice wafer stove are likely different than the Mayon Turbo stove, the amount of carbon emitted to the atmosphere is also likely different. Therefore, the amount of carbon dioxide emitted is calculated as in equation (5.2), from a modified version of equation (5.1):

$$CO_2 = C_e / [X_{CO_2} + [(CO:CO_2)*X_{CO} + (CH_4:CO_2)*X_{CH_4} + (EC:CO_2)*X_{EC} + (OC:CO_2)*X_{OC} + (NMHC:CO_2)*X_{NMHC} + (CH_2O:CO_2)*XCH_2O]] \qquad (5.2)$$

Where C_e is carbon emitted, X_Y is the molar mass ratio of carbon to compound Y, $CO:CO_2$ is the mass ratio of CO to carbon dioxide, EC is elemental carbon, OC is organic carbon, NMHC is nonmethane hydrocarbons, and CH_2O is formaldehyde. As 40 percent of the biochar is carbon by weight, then for each tonne of rice husk feedstock 121 kilograms of carbon is in the biochar and 279 kilograms of carbon is emitted, meaning that 30 percent of the carbon is retained in the biochar during cooking.

Table 5.5 Calculated Air Emissions from Rice Wafer Stove per Tonne of Biochar Produced

	Emissions (kg per tonne of biochar)
Carbon dioxide	2.59
Carbon monoxide	0.28
Methane	0.0068
Elemental carbon	0.0010
Organic carbon	0.0018
Nonmethane hydrocarbons	0.0105
Formaldehyde	0.0018
Nitrous oxide	0.0003
Nitrogen oxides	0.0022

Source: World Bank.

Unfortunately, the MacCarty et al. (2007) study was unable to measure the elemental carbon and organic carbon emissions for the Mayon Turbo stove, but provides emission factors for elemental and organic carbon for five other stoves (three-stone fire, rocket stove, Karve gasifier, Philips prototype fan stove, and jiko charcoal stove). The average emissions of elemental and organic carbon from these other stoves are used. From equation (5.2) and the emission ratios from MacCarty et al. 2007, the baseline emissions data are calculated as in table 5.5.

Economic Allocation of Impacts Associated with Cooking vs. Biochar Production
In order to differentiate between the impacts associated with biochar production and those from rice wafer production, economic allocation is used to weight the appropriate upstream impacts, calculated in the following manner. Each rice wafer takes approximately 30 seconds to cook. Cooking for 12 hours per day at 80 percent capacity, approximately 1,152 wafers per day are made. The value of each rice wafer is 500 VND,[18] or $0.02. Thus, the value of rice wafers produced is about $26.18 per day. The biochar value per day is $0.13, using a biochar price of 4,000 VND per bag, or $3.64 per tonne of wet biochar and assuming the biochar is sold by wet weight. The total income per day per stove is $26.31. The value of the biochar is 0.6 percent of the total value, while the rice wafers are 99.4 percent of the total. Therefore, less than 1 percent of the economic and GHG impacts of the following upstream processes are included in the LCA: cooking, stove construction, rice husk feedstock, and avoided rice husk burning.

Transport
The approximate cost to transport goods (the rice husk mesh bags or the bagged biochar) by bullock cart is 40,000–50,000 VND per tonne per kilometer.[19] A median value of $2.05 per tonne-kilometer is used. No environmental impacts are associated with using the bullock cart.

Biochar Soil Application
On the peanut farms, the biochar was applied two times, at a rate of 10 tonnes of wet biochar per hectare, for a total application of 20 tonnes of wet biochar per

hectare.[20] Assuming a 40 percent moisture content (wet basis),[21] this corresponds to 12 tonnes of dry biochar per hectare. Using an average carbon content of rice husk biochar of 40 percent (Haefele et al. 2009; Slavich et al. 2010), the application rate is equivalent to 4.8 tonnes of carbon per hectare. The biochar is applied by hand, as no machinery is available for this purpose.

The difference in labor costs between the biochar-amended plots and the nonbiochar-amended plots is calculated based on the difference in labor required for weeding, irrigating, and fertilizing.[22] For the biochar-amended plots, less labor is required for weeding and irrigation (savings of $19 and $13 per hectare, respectively), while more labor is required for fertilizing (cost of $3 per hectare to apply the biochar). The labor is estimated at 60,000 VND per hour, or $2.72 per hour. Thus, the net labor comes to −$30 per hectare, or −$1 per tonne of dry feedstock.

Carbon in the Biochar
The rice husk biochar has an average carbon content of 40 percent by weight (Haefele et al. 2009). Similar to the Kenya cookstove case study, the baseline analysis uses a conservative estimate of 80 percent recalcitrant carbon (Baldock and Smernik 2002; Lehmann et al. 2009), and the remaining 20 percent of the carbon is emitted as carbon dioxide in the short term.

Biochar Effect on Peanut Productivity
Peanut crops are grown twice per year in the study region. There have been three biochar trials on peanut crops, and the fourth crop trial is in progress. The experiment is laid out in a random complete block design with three replicates, a plot size of 10 square meters (2 meters × 5 meters), and plant density of 40 plants per square meter. Yield data for two of these crops (the two crops from 2009) are available for the project. The eight treatments were: no amendments, manure, NPK (nitrogen, phosphorus, and potassium) fertilizer, manure + NPK, biochar only, biochar + manure, biochar + NPK, and biochar + manure + NPK. The yield data provided by Slavich et al. (2010)[23] are presented in table 5.6. The manure application rate is 5 tonnes per hectare and the biochar application rate is 20 tonnes of wet biochar per hectare. These were not varied either in combination or individual applications. The fertilizer is applied at 30 kilograms of nitrogen per hectare, 26 kilograms of phosphorus per hectare, and 75 kilograms of potassium per hectare, in the form of urea, superphosphate, and potassium chloride, respectively (Slavich et al. 2010).[24]

According to the data available thus far and correspondence with the project contacts, the highest yield treatment is with the biochar + manure + NPK, with almost as much benefit from biochar + NPK.[25] The value of manure is being questioned as it is very expensive, and the results indicate that the biochar could substitute for the manure. According to Tam,[26] farmers can afford the full fertilization (see table 5.9 for prices). Thus, in deciding which treatments to utilize for the baseline scenario, correspondence with the project contacts has indicated that farmers would continue the NPK fertilization and potentially replace the

Table 5.6 Peanut Yield Data for Different Treatments

Treatment	Peanut yield (tonnes per hectare)		
	Crop 1	Crop 2	Average
None	1.17	0.99	1.08
Biochar	1.51	1.66	1.59
NPK	1.68	1.53	1.61
Biochar + NPK	2.21	1.88	2.05
Manure	1.38	1.58	1.48
Biochar + manure	1.79	1.66	1.73
Manure + NPK	1.93	1.60	1.77
Biochar + NPK + manure	2.57	2.00	2.29

Sources: Slavich et al. 2010; P. Slavich, survey response, 2010.
Note: NPK = nitrogen, phosphorus, and potassium (fertilizer).

manure with biochar, which is less expensive and results thus far show it to be longer lasting.[27] Thus, the control yield for the LCA is the manure + NPK treatment (1.77 tonnes per hectare), and the baseline yield used is the biochar + NPK treatment (2.05 tonnes per hectare), highlighted grey in table 5.6. The difference between the manure + NPK control treatment and the biochar + NPK baseline treatment is statistically significant, where this difference is 0.28 tonnes per hectare, or a 16% yield increase. The sensitivity analysis considers a range in crop yield effect from −50% to +47%, which allows for exploring the variability in the extent of the preexisting soil degradation, farmer practices, weather conditions, and fertilization rates. The maximum value of +47% increase is based on the statistically significant difference between the no amendment and biochar only treatments.

The surplus peanuts would likely be sold on the market. The price of peanuts is between 15,000 and 18,000 VND per tonne.[28] The median value of 16,500 VND, or $750 per tonne, is used for the baseline, and the sensitivity analysis explores this range.

Duration of Biochar's Agronomic Effect
The interior floodplains and the coastal zone have very sandy soils, which result in poor nutrient and water retention. The heavy rains in the wet season leach nutrients, while in the dry season the sand does not hold sufficient water and nutrient uptake is also restricted by a lack of adequate water. Preliminary results have found that biochar's water-holding capacity is 2 grams per gram, while the sandy soil alone is only 0.1 grams per gram.[29] Although data are not yet available, correspondence with project contacts has indicated that the soil constraints that the biochar addresses are the water- and nutrient-holding capacities, both of which are expected to be long-term effects and last for the lifetime of the biochar in the soil. As discussed for the Kenya case study, the duration of biochar's effect on soil properties and crop productivity is an important parameter in quantifying

the life-cycle impacts of biochar production and use. The baseline scenario estimates a 50-year agronomic effect for the applied biochar, while a "per crop" effect is also considered. The sensitivity analysis investigates this aspect further.

Soil Organic Carbon from Residue Removal

Although the removal of crop residues such as corn stover, rice straw, and rice husks would decrease soil organic carbon on the field from which they were removed, this soil organic carbon depletion is not included in this case study because of the nature of the feedstock. Even without rice wafer cooking or biochar production, the rice husks would be removed from the field with the rice grain.

Soil Organic Carbon from Increased Productivity

No data were available on changes in soil organic carbon from peanut crop residues. For this reason, the soil organic carbon to crop residue ratio for corn stover is used (as described in the Kenya case study). This estimate of the annual increase in soil organic carbon with stover left on the field is 0.01 tonnes of carbon per tonne of residue. Field measurements from the project demonstrate the difference in aboveground biomass with the different treatments (table 5.7).[30] It is common practice in the project region to remove the aboveground residues after the peanut harvest and use these for animal feed. For this reason, changes in soil organic carbon are assumed to come only from increases in belowground biomass. An average root to shoot ratio of 0.52 for Virginia-type peanuts (Huang and Ketring 1987) is used to estimate the change in belowground biomass. Similar to the change in crop productivity, the baseline scenario assumes the biochar + NPK treatment, as compared to the manure + NPK treatment (both highlighted grey in table 5.7). From the calculated belowground biomass, the increase in belowground residue yield between these two treatments is +0.23 tonnes per hectare per peanut crop. Using the 0.01 tonnes of carbon per tonne of residue rate, then it is estimated that +0.0023 tonnes of carbon per hectare is accumulated as soil organic carbon per peanut crop with the biochar + NPK

Table 5.7 Aboveground Biomass Residue Yield per Peanut Crop

Treatment	Total aboveground biomass (tonnes dry matter/hectare)
None	2.46
Biochar	3.56
NPK	3.33
Biochar + NPK	4.00
Manure	3.18
Biochar + manure	3.63
Manure + NPK	3.56
Biochar + NPK + manure	4.59

Source: Data from H.M. Tam, personal communication, 2011.
Note: The total is the sum of the pod and aboveground biomass. NPK = nitrogen, phosphorus, and potassium (fertilizer).

scenario. For two peanut crops per year and a 50-year biochar effect, the soil organic carbon contributes −70 kilograms of CO_2e per tonne of dry biochar in GHG reductions.

Avoided Manure

The GHG impacts of manure production are not included because the manure is produced regardless of biochar application. However, the avoided cost of the manure is included, as well as the avoided transportation costs of the manure. The price of manure in the region is around $18–23 per tonne, where the median value in this range is used. The avoided manure transportation is assumed to be 5 kilometers.

Fertilizer Use Efficiency

Measurements have been made on the nitrogen, phosphorus, and potassium uptake for all eight treatments. Results from the biochar + NPK and NPK only treatments are shown in table 5.8. Differences in NPK and biochar treatments are statistically significant for phosphorus and potassium uptake ($p = 0.029$ and 0.001, respectively), but not for nitrogen uptake. Therefore, the LCA considers the effect of improved fertilizer use efficiency with biochar additions for phosphorus and potassium only, where the uptake increases by 17.2 percent and 22.0 percent, respectively, for the biochar + NPK treatment as compared to the application of fertilizer alone. Literature values for improved fertilizer uptake are dependent on the crop, soil type, fertilizer application rate, and biochar application rate, but improved uptake values of up to 74 percent have been found for nitrogen when biochar was applied (Chan et al. 2007; van Zwieten, Kimber, Downie et al. 2010). As the farmers use full fertilization, it is assumed that the improved nutrient uptake and fertilizer use efficiency means less chemical

Table 5.8 Nitrogen, Phosphorus, and Potassium Uptake of Peanuts as Result of Mineral Fertilizer Applications of NPK

	Nutrient uptake (kilograms per hectare)		Difference
	NPK + biochar	NPK only	(%)
Nitrogen	22.8	20.1	9.0
Phosphorus	15.1	10.6	17.2
Potassium	38.8	22.4	22.0
	Nutrient recovery (%)		
Nitrogen	76.0	67.0	
Phosphorus	57.7	40.5	
Potassium	51.9	30.0	

Source: H.M. Tam, personal communication and unpublished data, 2011.
Note: Table shows nitrogen, phosphorus, and potassium uptake of peanuts as a result of mineral fertilizer applications of NPK at 30, 26, and 75 kilograms per hectare, respectively; recovery is calculated on a mass base in comparison to no additions of NPK. Differences in NPK and biochar treatments are statistically significant for phosphorus and potassium uptake.

Table 5.9 Fertilizer Prices in Vietnam

Fertilizer	Price (VND per kg)	Price	Unit
Urea	9,500	0.43	$ per kg urea
46-0-0		0.94	$ per kg nitrogen
Superphosphate	2,500	0.11	$ per kg superphosphate
0-18-0		0.63	$ per kg phosphorus
Potassium chloride (KCl)	11,000	0.50	$ per kg KCl
0-0-60		0.83	$ per kg potassium

Source: H.M. Tam, personal communication and unpublished data, 2011.
Note: VND = Vietnamese dong.

fertilizer need be applied to obtain the same yields. Thus, with biochar application, there are 17.2 percent and 22.0 percent savings in phosphorus and potassium, respectively.

Soil Nitrous Oxide and Methane Emissions
Similar to the Kenya case study, the baseline scenario of this analysis assumes that there are no changes in soil nitrous oxide or methane emissions with biochar application, because of the lack of site-specific data. However, the sensitivity analysis considers a range of +50% to –50% for the effect of biochar application on soil nitrous oxide emissions. (See the Kenya case study and the section titled "Impacts on Climate Change" in chapter 3 for further discussion on soil nitrous oxide emissions.)

NPK Fertilizer
The NPK fertilizer application rates are as in table 5.8. Data on the production of NPK fertilizers used in Vietnam were not available. Therefore, as a baseline, U.S. data are used, taken from the Greenhouse Gases, Regulated Emissions, and Energy Use in Transportation (GREET) 1.8b model (Wang 2007). However, due to lack of data, the transportation of these fertilizers to the field is not included. The fertilizer prices are as listed in table 5.9. If transportation of the fertilizers was included, the GHG reductions and monetary savings would be further increased.

Results and Discussion
Table 5.10 lists the result vector for the Vietnam rice husk biochar system. The net GHG balance is –0.5 tonnes of CO_2e per tonne of dry rice husk, as illustrated in figure 5.10. The net economic balance is +$948 per tonne of dry matter for the 50-year biochar effect. The majority of both the GHG emissions and reductions are from carbon dioxide, with much smaller contributions from methane and nitrous oxide (table 5.10). The surplus peanuts from biochar additions are 0.82 tonnes of peanuts per tonne of dry matter over the 50 years of biochar's effect on crop yields, or 16 kilograms of surplus peanuts per year.

Table 5.10 Result Vector for Vietnam Rice Husk Biochar System Baseline Scenario, per Tonne of Dry Rice Husk Feedstock

Input/output	Value	Unit
Methane	−0.13	kg
Nitrous oxide	−0.01	kg
Carbon dioxide	−509	kg
Stable carbon	113	kg
Net CO$_2$e	**−512**	**kg**
Revenue	+957	$
Cost	−9	$
Net $	**+948**	**$**
Surplus peanuts	0.82	tonne

Source: World Bank.
Note: For methane, nitrous oxide, and carbon dioxide, a negative sign corresponds to avoided emissions or sequestration, while a positive value indicates emissions. Negative $ are costs, and positive $ are net revenues. The bold type indicates the net GHG and economic balances. CO$_2$e = carbon dioxide equivalent.

Looking at the contribution analysis in figure 5.10, the largest source of reduced GHG emissions is the stable carbon in the biochar, at −0.4 tonnes of CO$_2$e per tonne of dry matter, or 81 percent of the total GHG reductions. The next largest amount of GHG reductions is from the avoided phosphorus and potassium fertilizer production, at 0.07 tonnes of CO$_2$e per tonne of dry matter, or 14 percent of GHG reductions. Soil organic carbon accumulation from increased belowground biomass contributes 5 percent to the GHG reductions, followed by the avoided rice husk burning at less than 1 percent. The only GHG emissions occur during rice wafer cooking and biochar production, which are very small compared to the net GHG balance (less than 1 percent). Another important aspect included in the analysis but not represented in figure 5.10 is that all of the rice husk feedstock is assumed to be renewable. Thus, only non-carbon dioxide GHG emissions are included in the stove emissions as these are assumed to not otherwise occur, whereas the carbon dioxide emissions are cancelled out by the uptake of carbon dioxide by the renewable feedstock carbon dioxide. If the feedstock were nonrenewable, then both carbon dioxide and non-carbon dioxide GHG would be included.

The contributions of the avoided rice husk burning and stove emissions to the net GHG balance are determined by the percentage allocation of these upstream processes. If the allocation for upstream processes is increased from 0.5 percent to 100 percent, the contribution analysis is as in figure 5.11. The GHG reductions increase to a net balance of −0.6 tonnes of CO$_2$e per tonne of dry matter. The contribution of the avoided burning is raised from −0.6 kilograms to −104 kilograms of CO$_2$e per tonne of dry matter, while the stove emissions increase from 0.5 kilograms to 9 kilograms of CO$_2$e per tonne of dry matter.

The −0.5 tonnes of CO$_2$e per tonne of dry matter obtained for the Vietnam rice husk biochar system is on the lower end compared to other life-cycle studies of different biochar systems in both developed and developing countries. Hammond

Figure 5.10 Contribution Analysis for Net Climate Change Impact per Tonne of Dry Feedstock for Rice Husk Biochar System in Vietnam

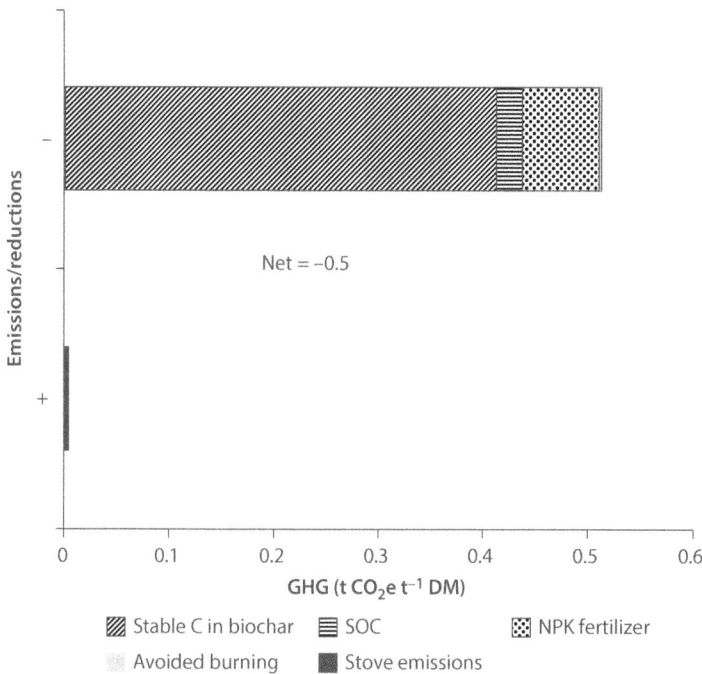

Net = −0.5

GHG (t CO_2e t^{-1} DM)

Emissions/reductions

▨ Stable C in biochar ☰ SOC ▨ NPK fertilizer
▨ Avoided burning ▪ Stove emissions

Source: World Bank.
Note: The upper bar (−) represents the GHG reductions, the lower bar (+) is GHG emissions, and the difference represents the net GHG balance of the system. C = carbon; SOC = soil organic compound; NPK = nitrogen, phosphorus, and potassium (fertilizer); GHG = greenhouse gas; CO_2e = carbon dioxide equivalent.

et al. (2011) obtained 0.7–1.3 tonnes of CO_2e per oven dry tonne of feedstock for slow pyrolysis biochar systems in the United Kingdom for small-, medium-, and large-scale process chains and 10 different feedstocks. Karve et al. (2011) found 0.9 tonnes of CO_2e per tonne of feedstock when assessing carbonized rice husks through gasification systems in Cambodia. Meanwhile, 0.8–0.9 tonnes of CO_2e per tonne of dry matter was found for crop residue and yard waste feedstocks for a large-scale pyrolysis system in the United States (Roberts et al. 2010).

The economic contribution analysis is presented in figure 5.12a for the base-line scenario of a 50-year biochar effect. The dry tonne of rice husk feedstock for biochar production and soil application results in $948 over the 50 years of biochar's agronomic effect. The majority of the revenue (75 percent) is from sales of surplus peanuts, while avoided manure use also contributes a significant portion (19 percent). Reduced phosphorus and potassium fertilizer needs, avoided transportation costs, and the biochar value contribute 5 percent, 1 percent, and 0.2 percent, respectively. In the figure, the transportation costs of the feedstock, biochar, and manure are aggregated, so that only the net transportation cost is represented, which is a revenue, because the avoided manure

Figure 5.11 Contribution Analysis for Net Climate Change Impact per Tonne of Dry Feedstock for Rice Husk Biochar System in Vietnam with 100 Percent Allocation for Rice Wafer Stove and Upstream Impacts

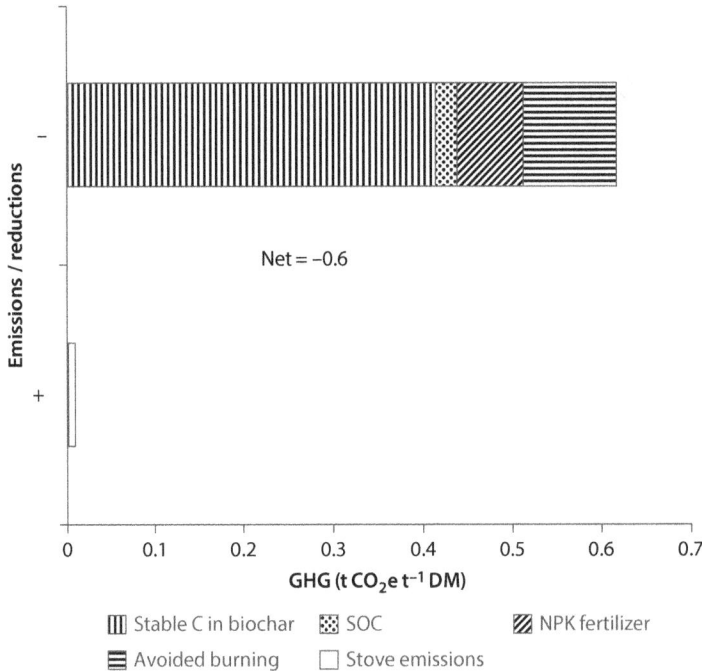

Source: World Bank.
Note: The upper bar (–) represents the GHG reductions, the lower bar (+) is GHG emissions, and the difference represents the net GHG balance of the system. C = carbon; SOC = soil organic compound; NPK = nitrogen, phosphorus, and potassium (fertilizer); GHG = greenhouse gas; CO_2e = carbon dioxide equivalent.

transportation revenue is higher than the feedstock and biochar transportation costs. The costs of the rice wafer feedstock and stove are small, again because of the economic allocation. If these impacts were allocated at 100 percent of biochar production, then the rice husk cost increases from $0.09 to $17 per tonne of dry matter and the stove from $0.002 to $0.40 per tonne of dry matter, for a net balance of $918 per tonne of dry matter.

A biochar "payback period" provides an estimate of the number of years it would take biochar used as a soil amendment to pay for itself. The payback period is calculated assuming the baseline price of peanuts, and is achieved in the first cropping season.

For a one-crop biochar effect, the results are presented in figure 5.12b. With the peanut surplus per crop only $7 per tonne of dry matter, the scale on the *x*-axis is greatly reduced. The transportation cost of the biochar becomes more apparent (–$6 per tonne of dry matter), and the avoided manure transport cost is minimal. The revenues from avoided manure and fertilizer are also decreased.

There is a distinction between the net revenue per functional unit and the net revenue for the farmer (table 5.11). For the farmer, the cost of the biochar incorporates the cost of the feedstock, the stove construction, and the biochar

Figure 5.12 Contribution Analysis for Net Economic Impact per Tonne of Dry Feedstock for Rice Husk Biochar System in Vietnam

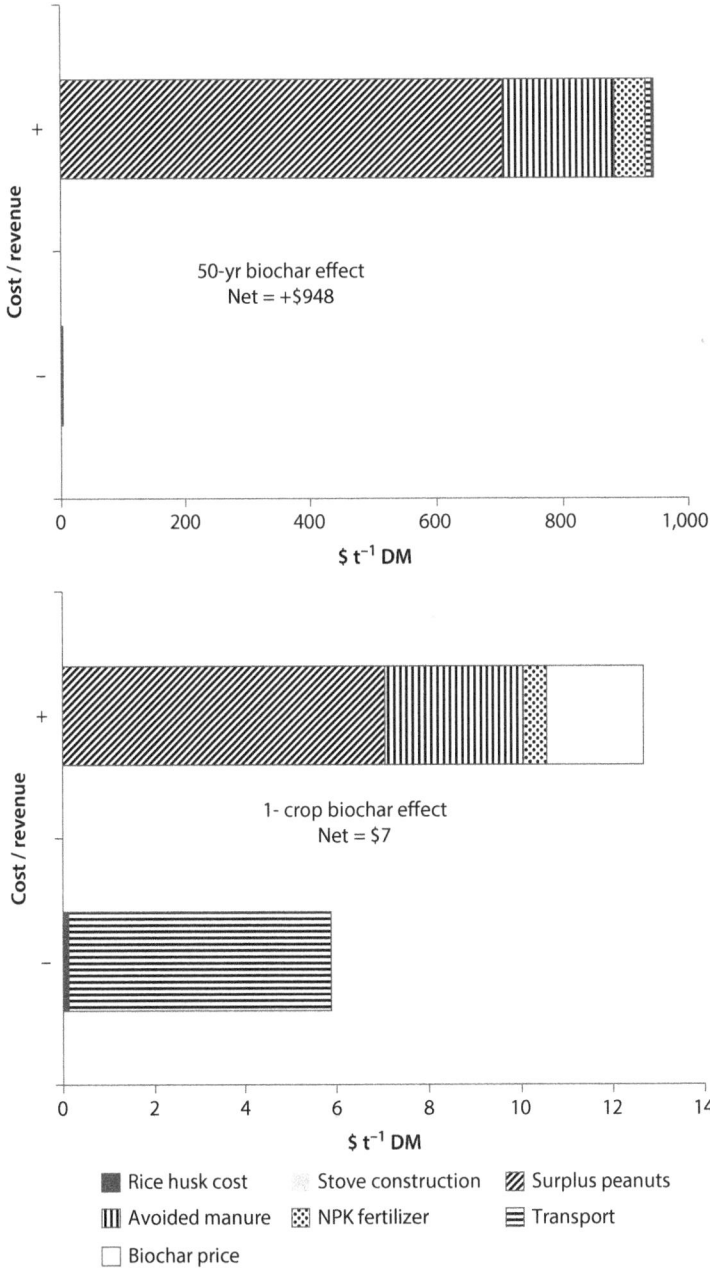

Source: World Bank.
Note: Figure 5.12a represents the baseline scenario, where biochar is assumed to have a 50-year agronomic effect; 5.12b is a scenario where biochar has only a one-crop effect. The upper bar (–) represents the revenues, the lower bar (+) is costs, and the difference represents the net economic balance of the system. NPK = nitrogen, phosphorus, and potassium (fertilizer).

Table 5.11 Comparison of Net Revenue between LCA Functional Unit and Farmer

	Costs and revenues ($ per tonne of dry matter)	
	Functional unit	Farmer
Feedstock	−0.09	
Stove construction	−0.002	
Biochar	2.14	−2.14
Surplus crops	704.72	704.72
Avoided manure	180.58	180.58
Rice husk transport	−0.06	−0.06
Biochar transport	−6.01	−6.01
Avoided manure transport	18.02	18.02
Avoided fertilizer	48.47	48.47
Total	947.75	943.59

Source: World Bank.

production, whereas for the functional unit, the biochar product is actually a revenue that offsets the cost of the stove, feedstock, and production. The other revenues from surplus crop sales, avoided manure and fertilizer inputs, and avoided transportation are the same per functional unit and per farmer. The difference in this case study is small ($4) because of the low cost of the biochar. However, in the case with a higher cost for biochar (and dependent on the specifics of the system under study), the economic balance per functional unit would be that much higher than for the farmer.

In this project, carbon credits are not received, but the farmer could potentially return an additional $13–15 per tonne of feedstock for a one-crop or 50-year effect and depending on whether only the stable carbon in the biochar or the life-cycle emission reductions receives offset credits. There is only a small difference in carbon credits between these scenarios because the GHG balance is dominated by the stable carbon in the biochar.

The $948 per tonne of dry matter achieved for the Vietnam rice husk biochar system is significantly higher than results from Roberts et al. (2010), where the most economically promising large-scale U.S.-based biochar system calculated in the analysis was yard waste at $69 per tonne of dry matter, which includes carbon offsets of $80 per tonne of CO_2e. However, the U.S. example assumed no increased crop productivity with biochar amendments, and only a one-crop biochar effect on maize yield. If there is no effect of biochar on the crop yields, then there are no surplus peanuts and thus no revenues are made from the sales of surplus peanuts. However, benefits from avoided manure and reduced fertilizer inputs could still be realized.

Sensitivity Analysis

A sensitivity analysis was conducted in order to measure the variability in the LCA results as a function of varying key input parameters (table 5.12). The input

Table 5.12 Sensitivity Analysis Input Parameters, Including Baseline and Range Values

Parameter	Baseline	Sensitivity range
Biomass throughput (tonnes of dry matter per year)	26	
Stable carbon content of biochar (%)	80	0–90
Yield response with biochar additions (%)	+16	−50 to +47
Peanut prices ($ per tonne)	750	682–818
Rice wafer stove methane emissions (kilograms per tonne of dry matter)	2.24	1.64–6.4
Avoided rice husk burning methane emissions (kg per tonne of dry matter)	3.71	3.71–25.7
Duration of biochar's effect (years)	50	1–100
Biochar transportation distance (km)	5	1–25
Soil nitrous oxide emissions (%)	0	−50 to +50

Source: World Bank.

parameters that were tested were the fraction of recalcitrant carbon in the biochar, the yield response of peanut crops with biochar additions, the price the farmer receives for peanuts, methane emissions of the rice wafer stove, methane emissions from avoided rice husk burning, the duration of biochar's agronomic effectiveness, the biochar transportation distance, and soil nitrous oxide emissions. The detailed sensitivity analysis is presented in appendix B, while a summary of the sensitivity analysis results is presented below.

The GHG balance is relatively insensitive (less than about 1 percent variability) to the rice wafer stove methane emissions, the rice husk-burning methane emissions, and the crop yield response with biochar additions. Within the realistic range of 50–90 percent, the fraction of the stable carbon

Figure 5.13 Sensitivity Analysis for Net GHG per Tonne of Dry Feedstock for Biochar System in Vietnam

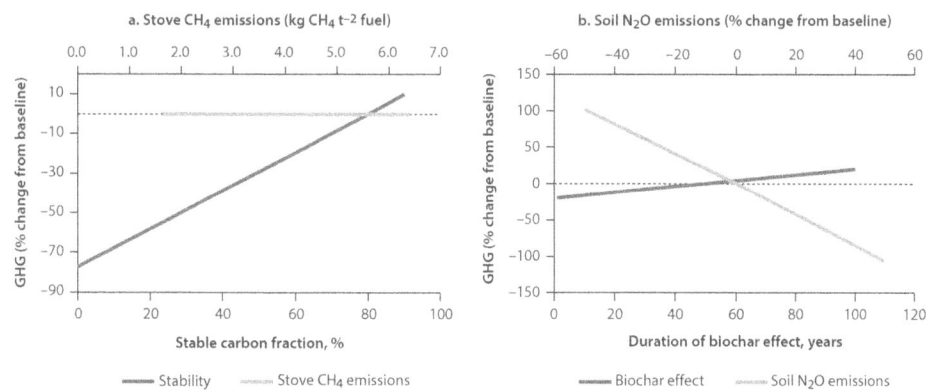

Source: World Bank.
Note: Figure 5.13a compares the fraction of stable carbon in the biochar to the rice wafer stove methane emissions; 5.13b illustrates the duration of biochar's agronomic effect and soil nitrous oxide emissions (nitrous oxide emissions range chosen according to extreme values available in the literature, see text). CH_4 = methane; N_2O = nitrous oxide; GHG = greenhouse gas.

Figure 5.14 Sensitivity Analysis for Net Economic Balance per Tonne of Dry Feedstock for Biochar System in Vietnam

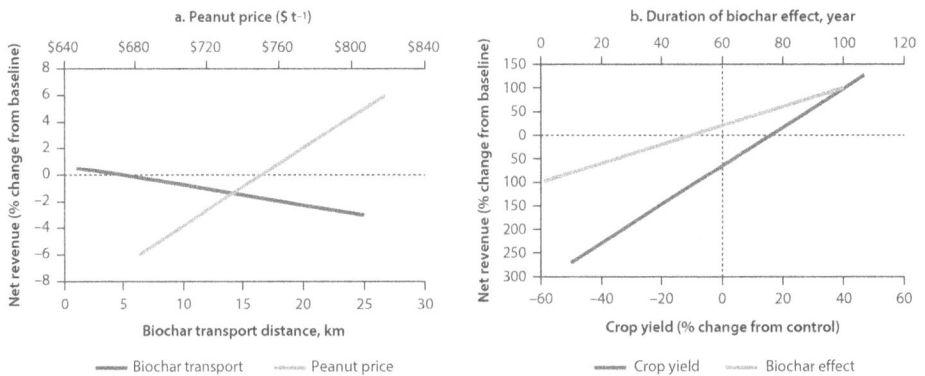

Source: World Bank.
Note: Figure 5.14a compares the biochar transportation distance to peanut prices; 5.14b illustrates the variation in the crop yield and the duration of biochar's agronomic effect.

in the biochar influences the net GHG balance by up to a 29 percent change from the baseline. Meanwhile, the GHG balance is most sensitive to the duration of biochar's effect (40 percent), and soil nitrous oxide emissions (100 percent). Figure 5.13 demonstrates that the net GHG balance for the Vietnam case study is most sensitive to the soil nitrous oxide emissions, followed by fraction of stable carbon in the biochar and the duration of biochar's effect, while the methane emissions during cooking have very little impact on the GHG balance.

Significant effects on the net economic balance are found within the range tested for the duration of biochar's agronomic effect (100 percent) and the crop yield response (260 percent), while the peanut price (6 percent) and biochar transportation distance (3 percent) have relatively small impacts. Figure 5.14 illustrates this variability, where it is evident that the crop yield response due to biochar additions, followed by the duration of biochar's effect, have the largest impact on the net economics. Meanwhile, the peanut price and the biochar transportation distance play only a small role in the economics of the system. More details on the sensitivity analysis can be found in appendix B.

Additional Considerations

Rice husk charcoal is currently used in traditional farming methods in some regions of Vietnam, thus there appears to be minimal cultural barriers to adoption of using biochar as a soil amendment. In addition, although quantitative data are not yet available, there is evidence of reduced irrigation needs on biochar-amended peanut plots as well. Reducing watering requirements is very important in this region of Vietnam, and is one of the primary goals of the overarching project.

Box 5.3 Summary of Vietnam Case Study

The Vietnam rice husk biochar system has potential for climate change adaptation and mitigation through carbon sequestration and GHG emission reductions, while also being economically viable for the smallholder farmer. The net GHG reductions are −0.5 tonnes of CO_2e per tonne of dry rice husk feedstock. The net economic balance is +$948 per tonne of dry matter over the 50 years of biochar's agronomic effect, or +$7 for a one-crop effect. In this project, carbon credits are not received, but the farmer could potentially return an additional $15 per tonne of feedstock with emissions trading at a price of $19 per tonne of CO_2e.

The contribution analysis reveals that the net GHG balance is largely (81 percent) driven by the stable carbon in the biochar, assuming that the rice husk remains a sustainable feedstock. Reduced fertilizer needs and soil carbon accumulation play lesser roles in the GHG balance. Soil nitrous oxide emissions could also play an important role in the GHG balance, thus more data are required to quantify this effect.

The sensitivity analysis reveals that the net GHG balance is relatively insensitive to changes in methane emissions during rice wafer cooking and avoided rice husk burning, the crop yield response with biochar additions, and the duration of biochar's agronomic effect, within the realistic range tested for these parameters. The net revenues are very sensitive to the crop yield response and the duration of biochar's agronomic effect, while the relative impact of the transportation distance of the biochar is less.

This Vietnam rice husk biochar system presents itself as a low-risk biochar project with climate change adaptation and economic benefits for the smallholder farmers incorporating biochar into their practices.

Senegal Case Study Life-Cycle Assessment

System Overview

The project is implemented by Pro-Natura International in Ross Bethio, Senegal. Charcoal and biochar are produced by a village-scale pyrolysis unit, the Pyro-6F, a continuous operation system. The original intent of the Pyro-6F was to produce "green charcoal," a renewable and cleanly produced fuel for household cooking. However, a growing interest in biochar as a soil amendment has increased the ratio of biochar to green charcoal production. The feedstock is rice husks obtained from a nearby rice mill, which would otherwise decay in piles. The biochar is purchased by farmers, and applied to onion, rice, and maize crops at a rate of 10 tonnes per hectare on farms in the area. The case study uses data from onions, as the onion data are the most mature at this time.

Methodology

Function, Functional Unit, and Reference Flows

For the Senegal case study, biochar production is a multioutput system with up to three coproducts: biomass management, soil amendment, and carbon sequestration. The functional unit of the system is 1 tonne of dry rice husk that is used as a feedstock at the pyrolysis unit, and the subsequent biochar is applied to onion fields on a farm in Senegal.

Figure 5.15 Schematic Flow Diagram for Rice Husk Biochar to Onion Production System in Senegal

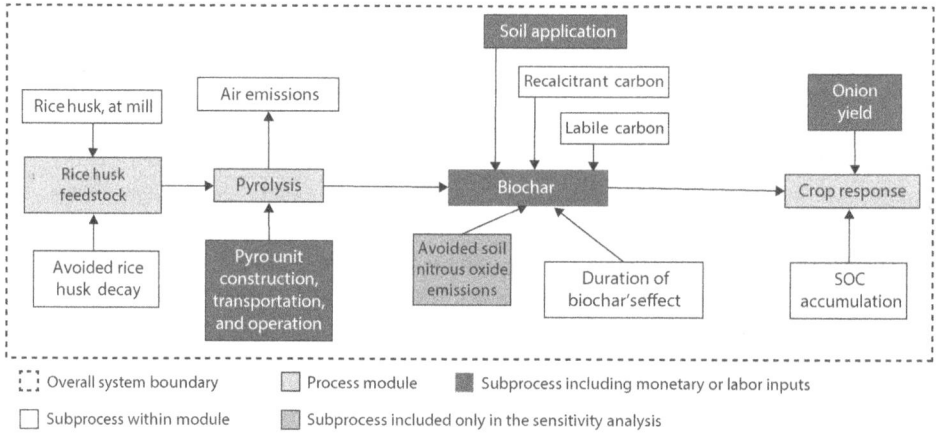

Source: World Bank.
Note: SOC = soil organic compound.

System Boundaries

A flow diagram of the biochar system is illustrated in figure 5.15. The system is organized into four modules: feedstock, pyrolysis, biochar, and crop response, each of which has multiple subprocesses.

Because the rice husk feedstock is a residue from the operation of an established rice mill, the production of the feedstock is considered a by-product and therefore no environmental burdens are associated with its generation (except for those impacts that would not otherwise occur in conventional management). However, if not utilized, the rice husk is piled up and left to decay. Thus, the avoided rice husk decay is included.

Under the pyrolysis module, the production and transportation of the unit are accounted for. During the operation of the unit, pyrolysis, labor and input requirements, and air emissions are included. The biochar is transported and applied to soils by hand. The behavior of biochar in soils is described by the stability of the carbon in the biochar. The recalcitrant and labile carbon fractions are included in the biochar module. The crop response upon biochar application is compared to yields for control crops. The effect of biochar on soil nitrous oxide emissions illustrated in the flow diagram is not in the baseline scenario but included in the sensitivity analysis. The system expansion method is used for modeling avoided rice husk decay.

Status of the Project

The Pyro-6F is an established system in Senegal and has been in operation by Pro-Natura since 2008, primarily making green charcoal. That same year, Pro-Natura launched a pilot project that provides biochar, training, and financial incentives to local farmers to facilitate the adoption of new sustainable agricultural practices based on biochar and organic fertilizers. The first biochar trials on vegetables showed encouraging results, and Pro-Natura has begun expanding trials to maize and rice on a larger scale. In addition, Pro-Natura has been working with Air France, through the intermediary of the Action Carbone program of the nongovernmental organization (NGO) GoodPlanet, to offer Air France passengers the option to compensate their carbon dioxide emissions with carbon credits generated primarily by the Pro-Natura green charcoal project in Senegal.

Life-Cycle Inventory

Rice Husk Feedstock

The project contacts state that rice hulls at mills are piled up outside the mill and left to decay. This allows a year-round supply of feedstock at no cost.[31] Thus, this LCA assumes that if the pyrolysis unit were not using the rice husks, then the rice husks would decay. Utilizing the rice husk as a fuel is also a waste management strategy.

Avoided Rice Husk Decay

Rice husks not utilized are piled up at rice mills and ultimately left to decay as a waste management strategy. The avoided methane emissions due to rice husk decay are calculated based on the Clean Development Mechanism (CDM) approved methodology for small-scale CDM project activities, category AMS-III.E, "Avoidance of methane production from decay of biomass through controlled combustion, gasification or mechanical/thermal treatment" (UNFCCC 2010). Because the rice husk is a waste stockpile, equation (5.3) in the CDM "Tool to determine methane emissions avoided from disposal of waste at a solid waste disposal site" is adjusted. The equation is as follows:

$$BE = \varphi \cdot (1 - f) \cdot GWP \cdot (1 - OX) \cdot CF \cdot F \cdot DOC_f \cdot MCF \cdot W \cdot DOC \cdot e^{-k(y-x)} \cdot (1 - e^{-k}) \quad (5.3)$$

Where φ is the model correction factor to account for model uncertainties (0.9), f is the fraction of methane captured and flared, combusted, or used in another manner (0), GWP is the global warming potential of methane (25), OX is the oxidation factor for the amount of methane oxidized in the soil or material covering the waste (0), CF is the molecular weight of methane relative to carbon (16/12), F is the fraction of methane in the gas (0.5), DOC_f is the fraction of degradable organic carbon than can decompose (0.5), W is the amount of organic waste prevented from disposal in the pile in the year x (1,388 tonnes), DOC is the fraction of degradable organic carbon by weight in the waste (0.4), k is the decay rate for the waste (0.01), x is the year during the crediting period (1), and y is the year for which methane emissions are calculated (2). Values from chapter 3, volume 5 of the 2006 IPCC Guidelines for National Greenhouse Gas Inventories (IPCC 2006) and the CDM III.E methodology are used for $\varphi, f, OX, F,$ and DOC_f. Both MCF and k are adapted for stockpiles of homogeneous wastes, at 0.284 and 0.01, respectively. W is calculated based on the approximate quantity of rice husk feedstock utilized by the Pyro-6F for biochar production per year. The result is that BE is calculated to be 11.5 tonnes of CO_2e per year, or 8.3 kilograms of CO_2e per tonne dry rice husk. Thus, the emission factor is 0.33 kilograms of methane per tonne of dry matter.

Figure 5.16 Photo and Diagram of Pyro-6F Unit

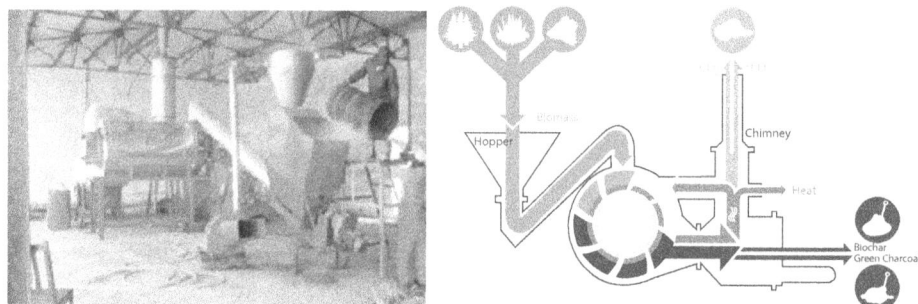

Sources: Figure 5.16a: Image of Pyro-6F in operation in Senegal (Pro-Natura International). Figure 5.16b: Schematic of Pyro-6F (Green Charcoal International).

Pyrolysis Unit Construction

The Pyro-6F unit (figure 5.16) is assembled in France and shipped via truck and barge to Senegal. The construction materials are estimated to be 3 tonnes of steel.[32] The GHG emissions associated with the steel production and transport are taken from the GREET 1.8b model (Wang 2007). The Pyro-6F is estimated to be in operation approximately 50 percent of the time.[33] The cost of the unit is $250,630, or $361,660. The capital cost is amortized over the life of the unit, where the equivalent annual capital cost is calculated based upon a 10-year lifetime and a 10 percent discount rate.

The truck transportation is estimated as the distances from Paris to Le Havre, France, at 196 kilometers, and from Dakar to Ross Bethio, Senegal, at 300 kilometers. The port-to-port distance from Le Havre to Dakar is approximately 4,360 kilometers.[34] The cost of transporting the unit is $10,000, or $14,430.[35]

Biochar Production

The Pyro-6F is a continuous pyrolysis furnace where the rice husk is heated to high temperature in a low-oxygen environment. The biomass is pyrolyzed into a carbon-rich material (biochar) and the gases are captured and combusted at high temperature to minimize products of incomplete combustion. The energy from combusting these gases is used to fuel the pyrolysis process.[36] Once running, the pyrolysis process is self-sustaining. However, 30–40 liters of gasoil is required at the start-up of the unit, and 7 kilowatts of electricity is consumed per hour for the continuous operation of the system. The biochar is cooled without coming in direct contact with water. It is then removed from the unit and bagged in 25-kilogram bags. Approximately 120 kilograms of biochar are produced per hour, with a biochar yield of 38 percent. Operating at 50 percent capacity, roughly 526 tonnes of biochar are produced per year, using 1,388 tonnes dry matter of rice husks per year. The rice husk feedstock is available at no cost from a nearby rice mill (200 meters away) and is transported by donkey cart. The biochar is sold to farmers for $200 per tonne.[37]

The operating cost of the Pyro-6F is estimated at $167 per year (assuming 50 percent operating capacity). This cost includes management, supervisor, operators, administration, fuel, electricity, packaging, and maintenance.[38]

Emissions Data

Emissions measurements for the Pyro-6F are not available at this stage. Because of the lack of pyrolyzer-specific data, emissions data from Brown 2009 are used for a controlled continuous charcoal kiln (table 5.13). As a conservative estimate the highest emission factors reported are used, at 8.9, 2.9, and 3.0 grams per kilogram of fuel for carbon monoxide, methane, and nonmethane hydrocarbons, respectively. As data on nitrous oxide emissions from the continuous kiln are lacking, a baseline value of 0.086 grams per kilogram of fuel is used, as calculated for the rice husk stove in the Vietnam case study. Variability in the emissions will be explored in the sensitivity analysis.

Transport, Donkey Cart

The cost to transport goods (the rice husk or the biochar) by donkey cart is estimated at $30 per tonne, from the difference in price of the biochar without transport ($200 per tonne) and biochar with transport ($230 per tonne).[39] Assuming an average transportation distance from the Pyro-6F to the farm of 10 kilometers, then cost of transporting is about $3 per tonne-kilometer. No environmental impacts are associated with using the donkey cart.

Biochar Soil Application

The biochar was applied at a rate of 10 tonnes per hectare to onion fields.[40] Using an average carbon content of rice husk biochar of 40 percent (Haefele et al. 2009), the application rate is equivalent to 4 tonnes of carbon per hectare. The biochar is applied by hand and takes the same amount of time as to prepare the field in the control, thus no additional labor requirements are included.

Carbon in the Biochar

The rice husk biochar has an average carbon content of 40 percent by weight (Haefele et al. 2009). The baseline analysis uses a conservative estimate of 80 percent recalcitrant carbon (Baldock and Smernik 2002; Lehmann et al. 2009), and the remaining 20 percent of the carbon is labile and emitted as carbon dioxide in the short term.

Table 5.13 Air Emissions Estimated for Pyro-6F per Tonne of Biochar Produced

	Emissions (kilograms per tonne of biochar)
Carbon monoxide	23.5
Methane	7.7
Nonmethane hydrocarbons	7.9
Nitrous oxide	0.0003

Source: World Bank.

Biochar Effect on Onion Productivity

Onion crops are grown twice per year in the study region. There has been only one biochar trial on onion crops, while other trials for maize and rice are in development. The experiment is laid out with three replicates and a plot size of 1 meter × 5 meters. The different treatments were:

- No amendments
- Compost (10 tonnes per hectare)
- NPK fertilizer
- Compost (10 tonnes per hectare) + NPK
- Biochar (5 tonnes per hectare)
- Biochar (5 tonnes per hectare) + compost (10 tonnes per hectare)
- Biochar (5 tonnes per hectare) + NPK
- Biochar (5 tonnes per hectare) + compost (10 tonnes per hectare) + NPK
- Biochar (10 tonnes per hectare)
- Biochar (10 tonnes per hectare) + compost (10 tonnes per hectare)
- Biochar (10 tonnes per hectare) + NPK
- Biochar (10 tonnes per hectare) + compost (10 tonnes per hectare) + NPK.

However, only the yield data for the biochar (10 tonnes per hectare) + compost (10 tonnes per hectare) + NPK, and its control of compost + NPK, are available. The biochar + compost + NPK results in 18.2 tonnes of onion per hectare (fresh weight), whereas the control is 12 tonnes per hectare, which is a yield increase of 52 percent.

The sensitivity analysis considers a range in crop yield effect from –50 percent to +233 percent, which allows for exploring the variability in the extent of the preexisting soil degradation, farmer practices, weather conditions, and fertilization rates. The maximum value of +233 percent increase is based on the difference between the biochar + compost + NPK versus compost + NPK for maize crop trials, but only relative yield data were available and not absolute values.

The surplus onions would likely be sold on the market. Few data are available for the local onion price. The estimated price the farmer receives is $287 per tonne, calculated based on a retail price of $552 per tonne in Senegal in 2002 (David-Benz, Wade, and Egg 2005) and a profit margin of 48 percent from farmer to retail (Weinberger and Pichop 2009) for African indigenous vegetables in Senegal. The sensitivity analysis explores the onion price and its impact on the net revenues of the system.

Duration of Biochar's Agronomic Effect

Although data are not yet available, the LCA assumes the constraints that the biochar addresses of these sandy soils in the project region are the water- and nutrient-holding capacity, both of which are expected to be long-term effects and last for the lifetime of the biochar in the soil. As discussed in the other case studies, the duration of biochar's effect on soil properties and crop productivity is an important parameter in quantifying the life-cycle impacts of biochar production and use. The baseline scenario estimates a 50-year agronomic effect for the applied biochar, while a "per crop" effect is also considered. The sensitivity analysis investigates this aspect further.

Soil Organic Carbon from Residue Removal

Although the removal of crop residues such as corn stover, rice straw, and rice husks would decrease soil organic carbon on the field from which they were removed, this soil organic carbon depletion is not included in the case study because of the nature of the feedstock. Even without biochar production the rice husks would be removed from the field with the rice grain.

Soil Organic Carbon from Increased Productivity

Data on changes in soil organic carbon from onion crop residues are unavailable at this time. For this reason, the soil organic carbon to crop residue ratio for corn stover is used (as described in the Kenya case study). This estimates that the annual increase in soil organic carbon with stover left on the field is 0.01 tonnes of carbon per tonne of residue. The LCA assumes that the aboveground residues are removed with the onion harvest, and thus changes in soil organic carbon are assumed to come only from increases in belowground biomass. An average bulb moisture content of 13 percent wet basis (Abhayawick et al. 2002), top to bulb ratio of 0.39 (Khan and Iortsuun 1989), and a shoot to root ratio of 2.81 (Azcón and Tobar 1998) are used to estimate the belowground biomass from the difference in onion bulb yields with and without biochar. From this method, the increase in belowground biomass is approximately 0.40 tonnes per hectare per onion crop. Using the 0.01 tonnes of carbon per tonne of residue rate, then it is estimated that an excess of +0.0041 tonnes of carbon per hectare is accumulated as soil organic carbon per onion crop with the biochar. For two onion crops per year and a 50-year biochar effect, the soil organic carbon contributes –146 kilograms of CO_2e per tonne of dry biochar in GHG reductions.

Fossil Fuel Production and Combustion

The emissions from fossil fuel production and combustion associated with truck and barge transportation, and residual oil production and combustion, are taken from the GREET 1.8b database due to lack of site-specific data (Wang 2007). The LCA assumes that the gasoil used for the Pyro-6F start-up is equivalent to residual oil from the GREET model and is combusted with similar emission factors to a utility boiler. The truck is assumed to be a class 8B diesel truck, and the barge consumes residual oil fuel. A gasoline generator provides the electricity consumed by the Pyro-6F. Emissions are based on a gasoline reciprocating engine from the GREET model, assuming a gasoline to electric generation efficiency of 35 percent.

Avoided Agricultural Inputs

Although some case studies have data on improved fertilizer use efficiency (Steiner et al. 2008; van Zwieten, Kimber, Downie et al. 2010) or reduced input needs, due to the lack of data for the Senegal case study the baseline scenario does not include any of these effects.

Soil Nitrous Oxide and Methane Emissions

Similar to the Kenya case study, the baseline scenario of this analysis assumes that there are no changes in soil nitrous oxide or methane emissions with biochar application, due to lack of site-specific data. However, the sensitivity analysis considers a range of +50 percent to –50 percent for the effect of biochar application on soil nitrous oxide emissions. (See Kenya case study and the section titled "Impacts on Climate Change" in chapter 3 for further discussion on soil nitrous oxide and methane emissions.)

Results and Discussion

Table 5.14 lists the result vector for the Senegal biochar system. The net GHG balance is –0.4 tonnes of CO_2e per tonne of dry rice husk, as illustrated in figure 5.17. The net economic balance is +$6,696 per tonne of dry matter for the 50-year biochar effect. The majority of both the GHG emissions and reductions are from carbon dioxide, with much smaller contributions from methane and nitrous oxide (table 5.14). The surplus onion from biochar additions is 23 tonnes over the 50 years of biochar's effect on crop yields.

Looking at the contribution analysis in figure 5.17, the largest source of reduced GHG emissions is the stable carbon in the biochar, at –0.4 tonnes of CO_2e per tonne of dry matter, or 87 percent of the total GHG reductions. The next largest amount of GHG reductions is from the soil organic carbon accumulation from increased belowground biomass, at –0.06 tonnes of CO_2e per tonne of dry matter, or 11 percent of GHG reductions. The avoided rice husk decay contributes 2 percent. The only GHG emissions are during the pyrolysis unit operation and transportation.

Another important aspect included in the analysis but not represented in figure 5.17 is that all of the rice husk feedstock is assumed to be renewable,

Table 5.14 Result Vector for Senegal Biochar System Baseline Scenario, per Tonne of Dry Rice Husk Feedstock

Input/output	Value	Unit
Methane	2.53	kg
Nitrous oxide	0.01	kg
Carbon dioxide	–480	kg
Stable carbon	121	kg
Net CO_2e	**–414**	**kg**
Revenue	6,815	$
Cost	–119	$
Net $	**6,696**	**$**
Surplus onion	23	tonne

Source: World Bank.
Note: For methane, nitrous oxide, and carbon dioxide, a negative sign corresponds to avoided emissions or sequestration, while a positive value indicates emissions. Negative $ are costs, and positive $ are net revenues. The bold type indicates the net GHG and economic balances. CO_2e = carbon dioxide equivalent

**Figure 5.17 Contribution Analysis for Net Climate Change Impact per Tonne of Dry
Feedstock for Rice Husk Biochar System in Senegal**

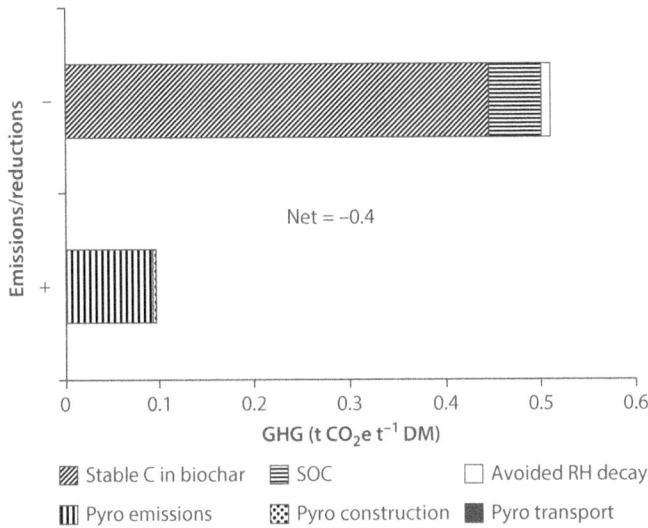

Source: World Bank.
Note: The upper bar (–) represents the GHG reductions, the lower bar (+) is GHG emissions, and the difference
represents the net GHG balance of the system. C = carbon; GHG = greenhouse gas; SOC = soil organic compound;
CO_2e = carbon dioxide equivalent.

similar to the Vietnam rice husk case study. Thus, only non-carbon dioxide GHG
emissions are included in the pyrolysis unit emissions as these are assumed to not
otherwise occur, whereas the carbon dioxide emissions are offset by the uptake
of carbon dioxide by the renewable feedstock. If the feedstock were nonrenewable, then both carbon dioxide and non-carbon dioxide GHG would be
included.

The 0.4 tonnes of CO_2e per tonne of dry matter obtained for the Senegal
case study is lower than other life-cycle studies in both developed and developing countries, which range from 0.7 to 1.3 (Hammond et al. 2011; Karve et al.
2011; Roberts et al. 2010). However, this lower value is expected, as the energy
is not captured and emissions from avoided fossil fuel combustion are not
included.

The economic contribution analysis is presented in figure 5.18a and 5.18b for
a 50-year and one-crop biochar effect, respectively. The dry tonne of rice husk
feedstock for biochar production and soil application results in $6,696 per tonne
of dry matter over the 50 years of biochar's agronomic effect, or $24 per tonne
of dry matter for a one-crop effect. The revenue comprises almost entirely (99
percent) the surplus onion sales, while biochar price adds a small amount. In the
figure, the transportation costs of the biochar and pyrolysis unit are aggregated
so that only the net transportation cost is represented. The capital and operating
costs of the pyrolysis unit are small in comparison to the revenues for the 50-year
effect, but can be seen in 5.18b and are significant compared to other case studies
(–$89 vs. –$1).

Figure 5.18 Contribution Analysis for Net Economic Impact per Tonne of Dry Feedstock for Rice Husk Biochar System in Senegal

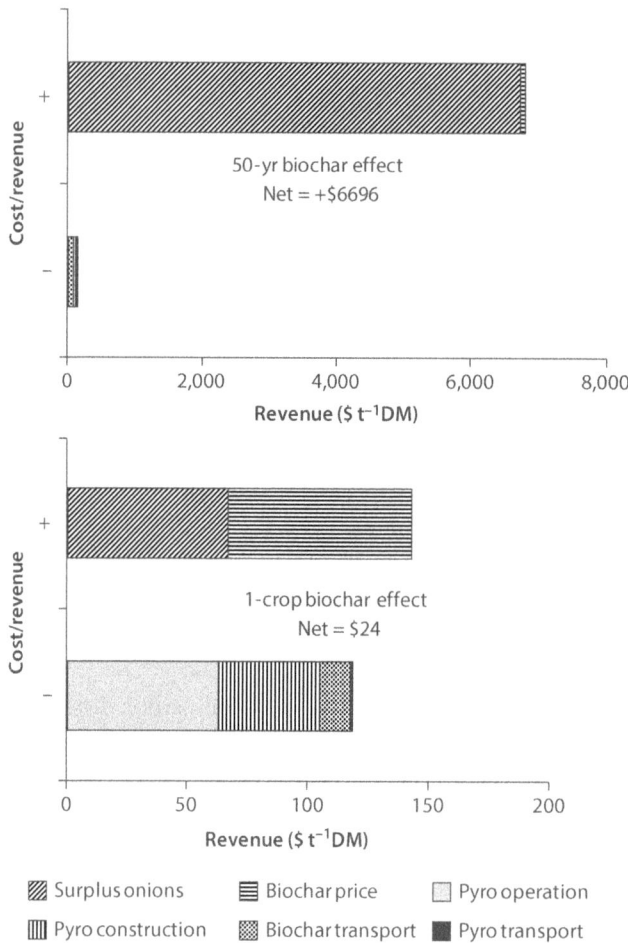

Source: World Bank.
Note: 5.18a represents the baseline scenario, where biochar is assumed to have a 50-year agronomic effect; 5.18b is a scenario where biochar has a one-crop effect. The upper bar (–) represents the revenues, the lower bar (+) is costs, and the difference represents the net economic balance of the system (note the different scales of both graphs).

There is a distinction between the net revenue per functional unit and the net revenue for the farmer (table 5.15). For the farmer, the cost of the biochar incorporates the cost of the feedstock, the pyrolysis unit capital, and operating costs, whereas per functional unit, the biochar product is actually a revenue that offsets the capital and operating expenses. The revenues from surplus crop sales are the same per functional unit and per farmer. The difference in net revenue per functional unit and per farmer is –$46. The economics can also be considered from the viewpoint of the NGO operating the Pyro-6F, as seen in the third column of table 5.15. Under the current parameters, the net economic balance results in a cost of –$30 per tonne of dry matter for the NGO. This comparison illustrates that despite the high potential for economic success for the farmer, the NGO

Table 5.15 Comparison of Net Revenue between LCA Functional Unit, Farmer, and NGO

	Costs and revenues ($ per tonne of dry matter)		
	Functional unit	Farmer	NGO
Kiln capital	−42	0	−42
Kiln operations	−63	0	−63
Kiln transport	−1	0	−1
Biochar	76	−76	76
Surplus crops	6,740	6,740	0
Biochar transport	−12	−12	0
Total	6,698	6,652	−30

Source: World Bank.

Note: NGO = nongovernmental organization.

might be operating at a loss. Depending on the funding scheme of the project, this could be problematic. The NGO might need to decrease operating costs (by increasing production time) or increase the price of biochar to the extent that farmers can still afford it.

In this project, carbon credits are not received, but the farmer could potentially return an additional $8 per tonne of feedstock if the stable carbon in the biochar receives offset credits for a carbon price of $19 per tonne of CO_2e.

The $6,696 per tonne of dry matter achieved for the Senegal biochar system is significantly higher than results from Roberts et al. (2010), where the most economically promising large-scale U.S.-based biochar system calculated in the analysis was yard waste at $69 per tonne of dry matter, which includes carbon offsets of $80 per tonne of CO_2e. However, the U.S. example assumed no increased crop productivity with biochar amendments, and only a one-crop biochar effect. If there is no crop productivity increase, there are no surplus crops, and thus no revenues from the sale of surplus crops.

Sensitivity Analysis

A sensitivity analysis was conducted in order to measure the variability in the LCA results as a function of varying key input parameters (table 5.16). The input parameters that were tested are the fraction of recalcitrant carbon in the biochar, the yield response of onion crops with biochar additions, the price the farmer receives for onions, methane emissions from the pyrolysis unit, methane emissions from avoided rice husk decay, the duration of biochar's agronomic effectiveness, the price of the biochar, the production time for the pyrolysis unit, the biochar transportation distance, and soil nitrous oxide emissions. The detailed sensitivity analysis is presented in appendix B, while a summary of the sensitivity analysis results is presented below.

The GHG balance is relatively insensitive to the Pyro-6F production time (less than about 1 percent variability) and the pyrolysis methane emissions (5 percent within the realistic range tested), and moderately sensitive to the

Table 5.16 Sensitivity Analysis Input Parameters, Including Baseline and Range Values

Parameter	Baseline	Sensitivity range
Biomass throughput (tonnes of dry matter per year)	1,388	
Stable carbon content of biochar (%)	80	0–90
Yield response with biochar additions (%)	52	−50 to +233
Onion price ($ per tonne)	287	70–552
Pyro-6F methane emissions (kg per tonne of dry matter)	2.9	2.2–6.4
Avoided rice husk decay methane emissions (kg per tonne of dry matter)	0.33	0.33–2.46
Duration of biochar's effect (years)	50	1–100
Biochar price ($ per tonne)	200	100–300
Pyro-6F production time (% of capacity)	50	25–100
Biochar transportation distance (km)	10	1–25
Soil nitrous oxide emissions (%)	0	−50% to +50

Source: World Bank.

duration of biochar's agronomic effect (15 percent) and rice husk decay methane emissions (15 percent). Meanwhile, the stable carbon content (15–40 percent) and soil nitrous oxide emissions (170 percent) could play the largest role in variability of the GHG balance. Figure 5.19 demonstrates that the net GHG balance for the Senegal case study may be most sensitive to the soil nitrous oxide emissions, followed by the fraction of stable carbon in the biochar and the duration of biochar's effect, while the methane emissions during pyrolysis have less impact on the GHG balance.

Significant effects on the net economic balance are found within the range tested for the crop yield response (up to 350 percent), the duration of biochar's agronomic effect (100 percent), and the onion price (up to 90 percent), while

Figure 5.19 Sensitivity Analysis for Net GHG per Tonne of Dry Feedstock for Biochar System in Senegal

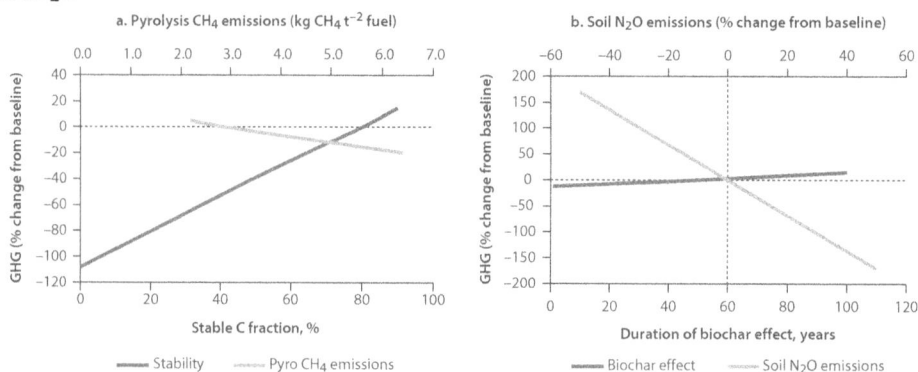

Source: World Bank.
Note: Figure 5.19a compares the fraction of stable carbon in the biochar to the Pyro-6F methane emissions, and 5.19b illustrates the duration of biochar's agronomic effect and soil nitrous oxide emissions (nitrous oxide emissions range chosen according to extreme values available in the literature, see text). CH_4 = methane; N_2O = nitrous oxide; GHG = greenhouse gas.

Figure 5.20 Sensitivity Analysis for Net Economic Balance per Tonne of Dry Feedstock for Biochar System in Senegal

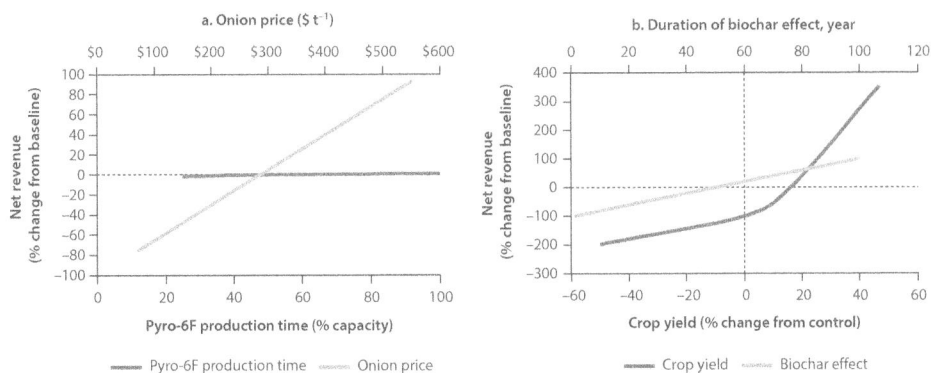

Source: World Bank.
Note: Figure 5.20a compares the onion price to the Pyro-6F production time; 5.20b illustrates the variation in the crop yield and the duration of biochar's agronomic effect.

the biochar price (3 percent), biochar transportation distance (less than 1 percent), and the Pyro-6F production time (2 percent) have relatively small impacts from the perspective of the functional unit. The price of surplus crops for the Kenya and Vietnam case studies had less impact on the net economics than for the Senegal study, though this was due to the high yield of onion crops with biochar additions and the resulting high revenues from surplus crops, leading to a stronger impact on the overall balance.

Figure 5.20 illustrates this variability, where it is evident that the crop yield response due to biochar additions followed by the duration of biochar's effect and the onion price have the largest impact on the net economics. Meanwhile, the Pyro-6F production time has only a small impact on the economics of the system, which is due to the fact that the net revenues are calculated per functional unit (1 tonne of dry feedstock). If the economic analysis were conducted from the viewpoint of the NGO operating the pyrolysis unit, the results would be different. From the NGO viewpoint, at 25 percent operating capacity the net revenues would decrease by around 350 percent, whereas increasing production to 100 percent would result in a net positive economic balance for the NGO (+$22 per tonne of dry matter) with an increase of around 170 percent. More details on the sensitivity analysis can be found in appendix B.

Box 5.4 Summary of Senegal Case Study

The Senegal biochar system has potential for climate change adaptation and mitigation through carbon sequestration and GHG emission reductions, while also being economically viable for the smallholder farmer. The net GHG reductions are −0.4 tonnes of CO_2e per tonne of feedstock. The net economic balance is +$6,696 per tonne of dry matter over the 50 years

box continues next page

Box 5.4 Summary of Senegal Case Study *(continued)*

of biochar's agronomic effect, or +$24 for a one-crop effect. In this project, carbon credits are not received, but the farmer could potentially return an additional $8 per tonne of feedstock with emissions trading at a price of $19 per tonne of CO_2e.

The contribution analysis reveals that the net GHG balance is largely (87 percent) driven by the stable carbon in the biochar, assuming that the feedstock remains a sustainable source. Soil nitrous oxide emissions could also play an important role in the GHG balance, thus more data are required to quantify this effect.

The sensitivity analysis reveals that the net GHG balance is relatively insensitive to changes in methane emissions during pyrolysis and avoided rice husk decay, the crop yield response with biochar additions, and the duration of biochar's agronomic effect, within the realistic range tested for these parameters. Soil carbon accumulation and transportation of the pyrolysis unit also play small roles in the GHG balance. The net revenues are extremely sensitive to the crop yield response, the duration of biochar's agronomic effect, and the price of onions, while the relative impact of the biochar transportation distance, the biochar price, and the pyrolysis unit production time is less.

This Senegal biochar system presents itself as a low-risk biochar project with climate change adaptation and potentially high economic benefits for smallholder farmers incorporating biochar into their practices.

Case Study Comparison and Conclusions

The net GHG balance of three biochar case studies in developing countries (Kenya, Vietnam, and Senegal) ranges from −0.4 to −1.8 tonnes of CO_2e per tonne of dry matter (figure 5.21). The Kenya cookstove project has the highest amount of GHG reductions due to the avoided emissions from traditional cooking (contributing −1.3 tonnes of CO_2e per tonne of dry matter), where 48 percent of the feedstock for traditional cooking would come from off-farm woody biomass, of which 80 percent is nonrenewable. The stable carbon in the biochar is the largest contributor to the GHG balance for the Vietnam and Senegal studies, and contributes about −0.4 to + 0.5 tonnes of CO_2e per tonne of dry matter for all three studies. The Vietnam case study highlights the role of reduced agricultural inputs (manure and phosphorus and potassium fertilizers) with biochar, further reducing GHG emissions. All systems analyzed demonstrate that the emissions from biochar production (whether cookstove or village-scale unit), transportation, and stove or kiln construction are minimal compared to the net balance of the system. Depending on the crop and increase in belowground biomass, increased soil organic carbon accumulation can also play a role, although small, in the net GHG of the system. Although not included in the baseline scenarios, the effect of biochar on soil nitrous oxide emissions can play a significant role in the net GHG of the system, emphasizing the need for improved data in this area.

In comparing the net revenues of the case studies, both the 50-year and one-crop biochar effect are discussed (figure 5.22). In figure 5.22, the surplus maize due to biochar additions in the Kenya study has been monetized for the discussion. The three case studies are presented in the chart side by side, but it is

Figure 5.21 Comparison of Net GHG Balance of Three Case Studies: Kenya, Vietnam, and Senegal

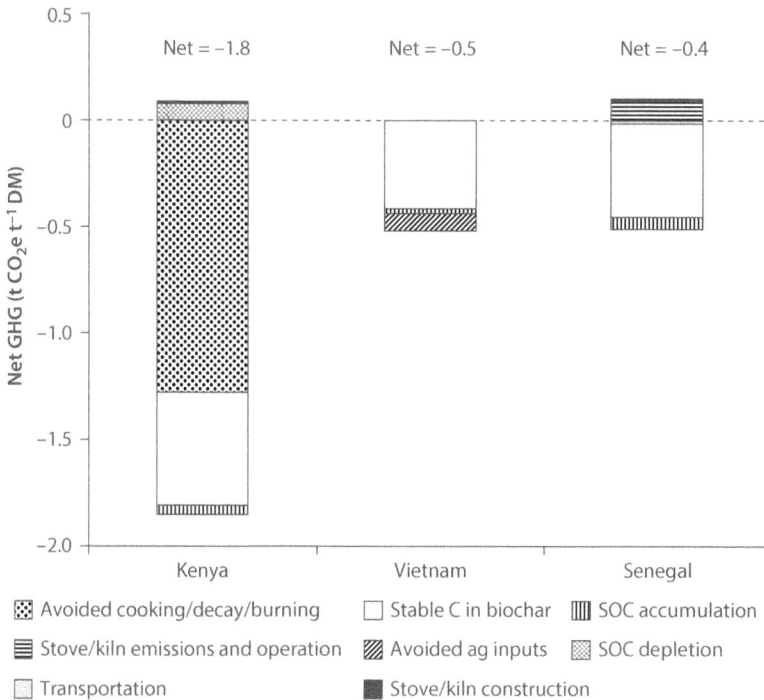

Source: World Bank.
Note: GHG = greenhouse gas; CO_2e = carbon dioxide equivalent; SOC = soil organic compound.

important to note that the comparison is not meant to select one project as better than another, as each project is unique with its own set of challenges and successes. Rather, the goal is to highlight strengths and weaknesses within each individual project, and the results are plotted alongside each other for convenience only. The most important result regarding the economics is that each project has a very short payback period—within one year when surplus crops are monetized. The Kenya case study demonstrates the added complexity of subsistence farming and valuing nonmonetary benefits such as increased food security for the household and labor savings.

The yield of the crops to which the biochar is applied plays the largest roles in determining the economic balance, implying that the farmer's choice of crops can be as important as the type of soil to which the biochar is applied. If biochar does not address local soil constraints, then crop productivity may not improve sufficiently (or at all) for farmers to realize an increase in revenues from sales of the surplus crops. For Senegal, the biochar is applied to a relatively high-yielding crop of onions. Although the price of onions is comparable to maize, the higher yield increase results in higher revenues for the farmer. Of course, the farmer's crop options may not be flexible, particularly in the short term.

Another important factor in the economic balance is the capital and operating costs of the biochar production technology. In the case of cookstoves, the capital

Figure 5.22 Comparison of Net Revenues of Three Case Studies: Kenya, Vietnam, and Senegal

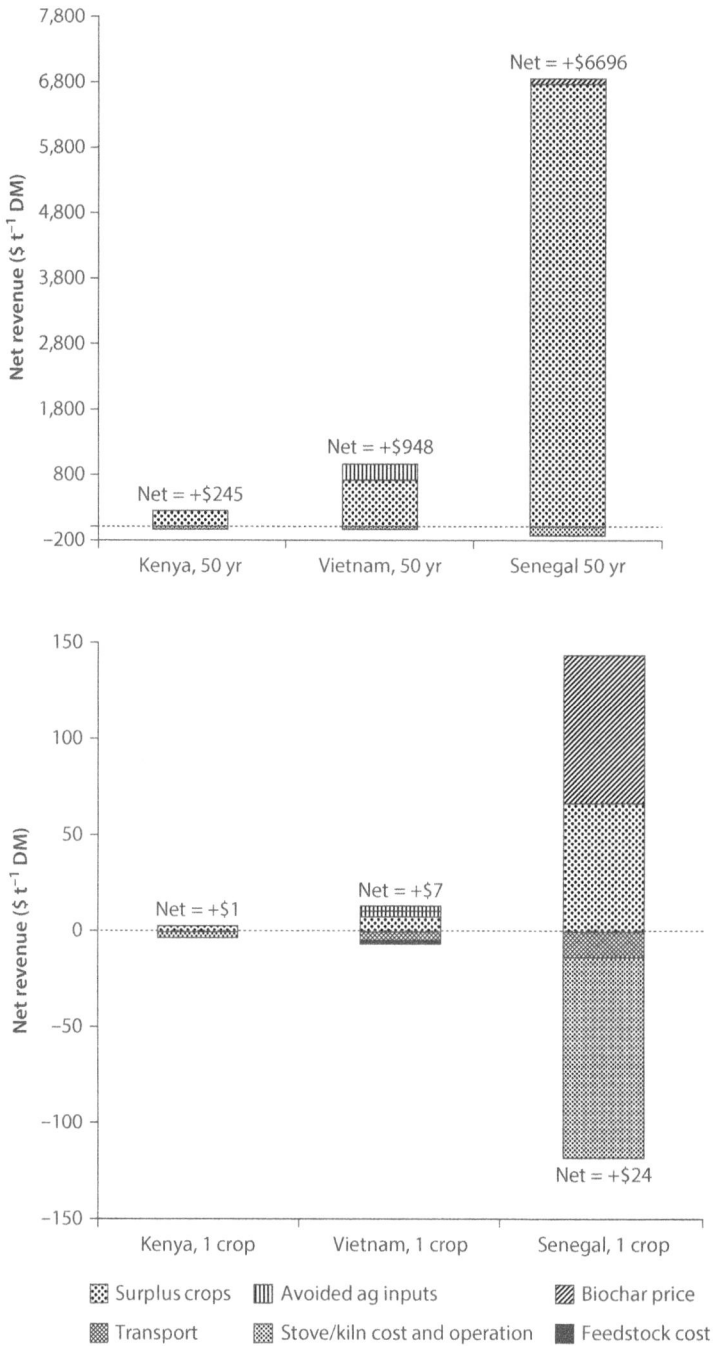

Source: World Bank.

Note: Figure 5.22a compares net revenues of the three case studies – Kenya, Vietnam, and Senegal – for a 50-year biochar effect; 5.22b compares net revenues for a one-crop biochar effect.

cost over the lifetime of the stove is small and the operating costs are minimal (cost of feedstock only). Meanwhile, the Senegal village-scale pyrolysis unit has significant costs, which are only offset by the large revenues from the surplus onion sales. If the biochar in Senegal were applied to maize, the same revenues might not be achieved because of the potentially lower yield increase of maize. The price the farmer receives for the surplus crop is also important for determining the economic balance, although less so than the crop yield response to biochar. In addition, the price of the biochar affects projects where the biochar is produced off-farm (Senegal and Vietnam). Avoided agricultural inputs and avoided transportation of those inputs are another source of revenue, as demonstrated by the Vietnam case study. For the LCA, the biochar price is considered a revenue per functional unit, which offsets the cost of the feedstock, capital, and operating costs, whereas the biochar is considered a cost to the farmer. Defining the success of the economic balance is dependent on the perspective, where the same costs and revenues may result in an economically favorable situation for the farmer but not for the NGO operating the pyrolysis facility, as in the case of the Senegal system.

The duration of biochar's agronomic effect plays a significant role in the economics of biochar systems in developing countries. On the high extreme, the Senegal study is calculated to generate +$6,696 per tonne of dry matter, assuming biochar has a 50-year effect, compared to +$24 per tonne of dry matter for a one-crop effect. However, it is important to note that even though a project may net +$948 per tonne of dry matter (as in the Vietnam case study, for example), this sum is distributed over the 50 years of biochar's effect. The farmer would achieve the surplus peanut revenue of $7 per tonne of dry matter each cropping cycle, in addition to reduced inputs and transportation costs each cropping cycle. Finally, the biochar application rate is also critical, where up to 950 percent variability in the net revenues of the system is found by decreasing the application rate an order of magnitude (Kenya case study). Determining the minimum biochar application rate that still achieves the desired agronomic response will enable farmers to make best use of limited biomass and economic resources.

Biochar projects in developing countries have the potential to reduce GHG emissions and be economically viable, as demonstrated by the life-cycle assessment case studies in Kenya, Vietnam, and Senegal. Ensuring the sustainability of the feedstock for biochar production is the first and most important step in achieving GHG reductions. With the feedstock sustainability in line, the recalcitrant carbon in the biochar is the largest source of direct carbon sequestration by removing atmospheric carbon dioxide and stabilizing it in the biochar. Avoided emissions from traditional biomass management practices such as traditional cooking can also play an important role where the feedstock is from a nonrenewable source, while avoided rice husk burning or decay are less influential because of the renewability of the resource. Emissions during pyrolysis (biochar production) have only a small impact on the net GHG balance for these systems. Meanwhile, the economics of these projects is largely dependent on the effectiveness of biochar to address soil fertility constraints, the duration of biochar's agronomic effect, the biochar application rate, and the value of the crops to which biochar is applied. Research and development efforts should focus on creating knowledge and understanding of these critical and interdependent parameters so that biochar projects in developing countries can be implemented with the highest probability for improving the livelihoods of smallholder farmers while also mitigating and adapting to climate change.

Notes

1. "The goal of the biochar energy, greenhouse gases, and economic (BEGGE) LCA is to quantify the energy, greenhouse gas, and economic flows associated with biochar production for a range of feedstocks" (Roberts et al. 2010).
2. D. Torres, unpublished data. It is possible that fuelwood gathering could be conducted at the same time as other chores, thus if multitasking were a common practice, the actual amount of time to collect fuelwood might be less.
3. J. Recha, personal communication, 2010.
4. Ibid.
5. Ibid.
6. J. Recha, personal communication, 2010.
7. J. Recha, personal communication, 2011.
8. D. Guerena, personal communication, 2010.
9. J. Lehmann, unpublished data, 2011.
10. Although this supplies the family with cooking energy for one year, the biochar effect is still assumed to be 50 years, as in the baseline scenario.
11. J. Recha, personal communication, 2011.
12. ICE website: Emissions https://www.theice.com/productguide/ProductGroupHierarchy.shtml?groupDetail=&group.groupId=19.
13. International Rice Research Institute website: Rice knowledge bank: rice husk http://www.knowledgebank.irri.org/rkb/index.php/rice-milling/byproducts-and-their-utilization/rice-husk.
14. P. Slavich, personal communication, 2011.
15. The Engineering Toolbox website: Estimating brick quantity and mortar consumption http://www.engineeringtoolbox.com/brick-mortar-consumption-d_1549.html.
16. P. Slavich, personal communication, 2011.
17. H. M. Tam, personal communication and unpublished data, 2011.
18. H. M. Tam, personal communication, 2011.
19. Ibid.
20. P. Slavich, survey response, 2010.
21. H. M. Tam, personal communication, 2011.
22. H. M. Tam, personal communication, 2011; P. Slavich, personal communication, 2011.
23. Also, P. Slavich, survey response, 2010.
24. H. M. Tam, personal communication, 2011.
25. P. Slavich, personal communication, 2011.
26. H. M. Tam, personal communication, 2011.

27. P. Slavich and H. M. Tam, personal communication, 2011.

28. H. M. Tam, personal communication, 2011.

29. P. Slavich, personal communication, 2011.

30. H. M. Tam, personal communication, 2011.

31. C. Braun, personal communication, 2011.

32. C. Braun and G. Reinauld, personal communication, 2011.

33. Ibid.

34. Data from Portworld http://www.portworld.com/map/.

35. C. Braun and G. Reinauld, personal communication, 2011.

36. Green Charcoal International website: Technology http://www.green-charcoal.com/pyrolyser.

37. C. Braun and G. Reinauld, personal communication, 2011.

38. Ibid.

39. C. Braun and G. Reinauld, personal communication, 2011.

40. Ibid.

CHAPTER 6

Aspects of Technology Adoption

Economics of Biochar

The results of the life-cycle assessment (LCA) case studies in chapter 5 and other biochar analyses (Gaunt and Cowie 2009; Gaunt and Lehmann 2008; Hammond et al. 2011; Roberts et al. 2010) demonstrate that biochar systems can offer potential greenhouse gas (GHG) reductions in the range of 0.4–2.0 tonne of carbon dioxide equivalent (CO_2e) per tonne of dry feedstock. However, the implementation and sustainability of all biochar systems, but particularly those in developing countries where start-up capital and other funds are lacking or severely limited, are dependent on the economics of these projects. If the economics of the projects are favorable to smallholder farmers, the adoption of biochar systems in smallholder agricultural systems will be facilitated and GHG emissions will be further reduced, while also allowing for adaptability and resilience to climate change.

The economics of biochar systems in developing countries are dependent on multiple factors that are specific to the project, and are further complicated in these systems by the fact that many benefits to the smallholder are not traditionally monetized. Examples of parameters that are more easily quantified include the cost of the feedstock (if any), the capital and operating costs of the stove or kiln, transportation of feedstocks and biochar, the price of biochar, the price (if any) of surplus crop yields, and the savings from reduced agricultural inputs and waste management. Meanwhile, parameters such as decreased labor, improved indoor air quality, and increased food for the household directly benefit the project participants but are not generally valued monetarily, particularly in developing-country settings. Furthermore, additional benefits may potentially arise due to decreased deforestation pressures, improved water use efficiency, and climate change adaptation. Finally, although none of the biochar projects to date receives carbon credits, this contribution may increase revenues as projects progress. Overall, the economics of biochar projects analyzed in the case studies are largely determined by the price farmers receive (or lack thereof) for surplus crops due to biochar additions.

Few economic analyses of biochar projects in developing countries have been published to date; the only publication at this time is Joseph 2009. Joseph estimates that for a hypothetical biochar project in a developing country using a combination of cookstoves and kilns, the net present value of a five-year project is $55,943, with 292 tonnes of biochar produced over the five years. Assuming a biochar yield of 22 percent as in this analysis, the net economic balance is +$42 per tonne of dry feedstock. This analysis includes aspects of implementing a development project not included in the LCA, such as extension and external consultants and compliance monitoring and evaluation, but also assumes carbon credits are received by the project (which the LCA in chapter 5 does not). The Kenya, Vietnam, and Senegal case studies calculate balances of –$1, +$955, and +$6,700 per tonne of dry feedstock, respectively. The wide range in economic balances is indicative of the unique, case-by-case nature of these types of projects. For example, subsistence farmers in Kenya may not see any economic return for increased maize sales, but the food supply for the family will be improved. If these supplies are valued monetarily, then the balance could be estimated at +$127 per tonne of feedstock.

Comparing the economic balances of the developing-country projects to biochar systems in industrialized countries, it is possible that the potential benefits for smallholder farmers may be greatest. For example, all the feedstocks analyzed by Roberts et al. (2010)—stover, yard waste, and switchgrass—required carbon credits in order to be profitable. On the other hand, the onion and peanut farmers in Senegal and Vietnam would potentially realize increased revenues within the first year of the project. Pratt and Moran (2010) also observe that small-scale, slow-pyrolysis biochar systems, such as those relying on cookstoves, are often more cost-effective for GHG reductions than the fast-pyrolysis systems favored by large biochar plants. One of the main causes of this is that biochar offers the most agronomic benefits on poor-quality soils, which is frequently the case for smallholder farmers in developing countries, as is exemplified by the case studies. The economics of biochar projects in developing countries may therefore be largely driven by the potential revenues from sales of surplus crops due to biochar additions to the soil.

Engagement with Carbon Markets

Criteria for a Successful Project

There are many factors that contribute to the success of a project designed to mitigate climate change by generating emission reductions, be it for the voluntary or compliance-based carbon markets. In a market scenario, the emission reductions generated by a project can be sold when their price is lower than the marginal cost of emission abatement the buyer faces. In a compliance market, these emission reductions are used directly as credits to allow an entity to emit a given amount of GHGs while remaining under the limit (cap) they have been allotted. In a voluntary market, carbon offsets are sold to companies, organizations, or individuals with the desire, but not the obligation, to reduce their carbon footprint. Thus, one can see why a more stringent level of assurance that emission reductions truly reduce GHG emissions as much as they are supposed to is needed for those sold within compliance-based markets.

A number of basic guidelines are required for a successful biochar project to reduce GHG emissions and to generate emission reductions. Whitman, Scholz, and Lehmann (2010) discuss these issues for biochar projects in greater detail, but here the following will be considered: additionality and baselines; permanence; leakage and system drivers; and measurement, reporting, and verification.

Additionality and Baselines

The principle of additionality requires that a project would not have taken place under a business as usual scenario and without the incentive provided by the price of carbon. The most prevalent tool used to determine additionality is the "additionality tool" used under the Kyoto Protocol's Clean Development Mechanism for generating carbon offsets (UNFCCC 2008). In order to successfully demonstrate additionality, the baseline scenario—what would have happened without the project in place—is critical. With biochar projects, the difference between the impact of creating biochar from biomass that would have been burned, grasses that decompose very quickly, slow-decomposing woody biomass, or a living tree is substantial and must be taken into account. If the baseline scenario was that the tree was going to continue growing, drawing carbon dioxide from the atmosphere, charring it would release so much carbon that, despite increased stability, the net carbon budget would register a carbon release to the atmosphere for decades, if not hundreds of years.

Permanence

The issue of permanence is discussed extensively in the section titled "Impacts on Climate Change" in chapter 3. While it is critical to establish what fraction of biochar will be relatively stable, the challenges associated with establishing and maintaining carbon storage with biochar are drastically different from those associated with nonbiochar soil carbon or forests. Carbon storage through nonbiochar soil carbon or forest management tends to be much more easily reversible, through changes in management practices. Once biochar is applied to soils, its impact depends more on its physical and chemical properties than on continuation of specific management practices.

Leakage and System Drivers

Leakage occurs when emission reductions within a project boundary result in increased emissions elsewhere (that is, outside the project boundary). For example, if forest conservation in one area resulted in increased logging in adjacent forests, this would constitute leakage. One robust way to try to prevent leakage is by taking a systems view, as is done in the LCA section of this report. Two examples of potential effects that may be overlooked in biochar projects include the potential for black carbon emissions to the atmosphere and biochar's effects on GHG emissions from soils. As discussed in the section titled "Impacts on Climate Change" in chapter 3, these effects are currently being studied. A second important source of leakage can come from "rebound effects." For example, if a family begins using a more efficient cookstove, if wood availability was a limiting factor before, they may cook more with the efficient stove. Thus, the predicted energy, carbon, and fuel savings may easily be overestimated. Some systems are more prone to this effect than others. The use of a cookstove is constrained somewhat by how much people eat, while the use of a system that provides energy for other uses would be less likely to be limited. Considering system drivers in combination with "true wastes" for feedstocks can help to create a "safer" biochar-producing system.

Measurement, Reporting, and Verification

In order to understand the climate effects of a project, it is essential to be able to quantify and verify its impact. The measurement of the impacts due to changes in energy use and production within a biochar system should not differ greatly from extant energy-based offset projects, nor should it necessarily produce many new challenges. The new challenge is in measuring the impact of the biochar. There are two ways one might think of its measurement—direct and indirect. Direct measurement—for example through repeated soil sampling and quantification—is attractive in some ways, such as its clear tangibility, but is challenging in others. Measuring biochar contents in soils would raise similar challenges to measuring nonbiochar soil carbon—its heterogeneity makes it difficult and expensive to measure. Furthermore, it would not be possible to capture biochar lost due to erosion from runoff, which has been found to provide a substantial proportion of the losses of biochar and fire-generated char from soils (Guggenberger et al. 2008; Major et al. 2010; Rumpel et al. 2006). Assuming any biochar removed from the system through runoff is equivalent to its return to the atmosphere would likely result in overly conservative estimates of its impact (Whitman, Scholz, and Lehmann 2010). An alternative to direct measurement might be to monitor its production, and, combined with decomposition studies, predict the persistence of biochar in soils, combined with minimal sampling to establish that the biochar is being applied to soils and not used in other ways. Such targeted verification measurements could use techniques to specifically test for biochar-type carbon forms (Manning and Lopez-Capel 2009), which would unambiguously prove the source of the soil carbon change. Developing confidence in this approach may be critical to enable biochar projects to include biochar-based emission reductions in their carbon accounting.

Future of Biochar in Carbon Markets

There are currently no approved methodologies under the Clean Development Mechanism (CDM) that include biochar. A submission of a methodology for a large-scale biochar production system has been made to the Verified Carbon Standard (formerly Voluntary Carbon Standard) (Carbon Gold 2009), but has not yet been approved. With regard to (nonbiochar) cookstoves, relevant methodologies already exist (ClimateCare 2010; UNFCCC 2009). Table 6.1 summarizes the state of existing protocols for measuring the various potential impacts of biochar systems (summarized earlier in table 3.2). Notably, what remains to be developed is a robust protocol for measuring the impact of the biochar itself.

Table 6.1 State of Protocols for Measuring GHG Impacts of Biochar Systems

Source	Protocols	Notes
Carbon stabilization	Carbon Gold submission to Verified Carbon Standard currently under review	The current submission is likely insufficient and is not based on strong scientific reasoning. Developing such a protocol is very important if biochar projects are to succeed in measuring their full climate change impact, as carbon stabilization in biochar comprises a substantial portion of emission reductions.
Renewable energy	Numerous CDM protocols; Gold Standard protocol for cookstoves	CDM protocols cover many aspects of the replacement of fossil fuel energy with renewable energy. There is no reason why adapting protocols for energy-related impacts to biochar projects would be particularly challenging.
Waste diversion	CDM methane avoidance	The CDM has a protocol for avoiding methane production of biomass through pyrolysis, limited to projects that produce less than 60,000 tonnes of CO_2e per year. Further development of similar protocols would be necessary, but must be synchronized with those that would count carbon stabilization in biochar to avoid double-counting of emission reductions.
Reduction in soil emissions	None	No protocol currently exists, and soil GHG emissions are notoriously challenging to measure due to high spatial and temporal variability.
Reduction in fertilizer manufacturing	None	No protocol is currently known to exist, but its quantification should not be particularly challenging for biochar projects as compared to others.
Increased non-BC^a soil carbon	None	While the sign of the impact of BC on soil carbon (positive/negative) should become predictable, its precise quantification would be extremely challenging and it is not likely to provide a major fraction of GHG reductions.

Source: World Bank.
Note: CDM = Clean Development Mechanism; CO_2e = carbon dioxide equivalent; GHG = greenhouse gas.
a. BC = black carbon.

The cost per tonne of CO_2e reductions predicted for an improved cookstove project in Mexico by Johnson et al. (2009) was around $8, which was in part due to the high installation costs of the stove itself (around $100). More recent experience in the field of improved cookstove projects shows that these costs can be brought down dramatically using economies of scale and by structuring the sale of the stoves accordingly.[1] With a stove cost of roughly between $4 and $30 per stove the cost per tonne of CO_2e reduction becomes much more attractive and in many instances might be very competitive to other abatement options. If a biochar cookstove in western Kenya were improved to produce emission reductions per stove of the same order of magnitude (Whitman et al. 2011), one might predict that the marginal cost of abatement might be similar to other improved cookstove projects. However, measurement and verification of the biochar production and application as potentially required by any future biochar methodology for climate change mitigation would likely add an additional cost.

Biochar Systems for Smallholders in Developing Countries • http://dx.doi.org/10.1596/978-0-8213-9525-7

Sociocultural Barriers to Adoption

No matter what the technical potential benefits of biochar are, their realization depends on whether and how people implement biochar systems. Understanding sociocultural barriers to adoption is essential for a successful project, but can be challenging and often requires highly location-specific understanding of people and their needs, values, and expectations. FAO (2010) provides an in-depth discussion of barriers and incentives to implementation common to food-energy systems. It notes that barriers to the implementation and wide-scale dissemination of such systems concern various aspects at both farm and beyond-farm levels. Key considerations include the following:

- The complexity of systems requires high levels of knowledge and skills.
- The technology used needs to be reliable and economical.
- Financing is mostly related to the investment required for the energy conversion equipment.
- The increased workload makes the systems less attractive to farmers.
- Competition between different uses of residues must be addressed.
- Access to markets for agricultural and energy products is often a key factor to ensure economic viability.
- Access to information, communication, and learning mechanisms is as important a production factor as "classic" land, labor, and capital.
- Few government policies encourage all aspects of systems.

Each item on this list could apply to biochar systems and the challenges of implementing them. This list could serve as a useful checklist for project developers and implementers.

Sociocultural factors can influence project success in both positive and negative ways. Potential impacts from the adoption of biochar systems were discussed in the section titled "Social Impacts" in chapter 3. These include, on the negative side, health impacts from biochar dust; increased labor to produce and apply biochar; increased cost to purchase or apply biochar; potential gender impacts; and competition for feedstocks needed for soil protection, fodder, fuel, or income. On the positive side, potential impacts include improvement to health through better nutrition from increased crop yields and substitution by clean cookstoves and kilns of less efficient, high-emission units; increased access to household energy, both thermal energy and, in some cases, electricity; and reduced labor and increased income through increased crop yields and reduced inputs of fertilizer and water. Which of these effects are the most important in a given scenario and region would likely influence the success of a given project.

To better understand the importance of some of these factors in current projects, a second survey was conducted, as described in the section titled "Survey" in chapter 4. This follow-up survey to the original survey asked respondents questions about barriers, traditional indigenous biochar use, how projects coped with limited supplies of biochar, the top perceived benefits, and project reliance on carbon financing. The survey included many open-ended questions, which allowed respondents to provide detailed responses.

The first set of questions asked directly about sociocultural barriers and ideas for overcoming them. Overwhelmingly, respondents cited a lack of awareness of biochar and the need for education and demonstration projects to show farmers that making and using biochar would be worth their time (see appendix A for documentation of results). Gender issues were perceived to be a minor barrier, making up 10 percent of the cited reasons for barriers, and one respondent thought biochar would help women by adding to their household income. Labor barriers were mentioned for slash-and-char systems, where biochar was in competition with slash-and-burn practices, which required less labor and gave a quicker result. The labor required to gather dispersed feedstocks was also cited as an obstacle. The availability of biochar production technology was a related barrier. Very few projects had yet developed an adequate supply of biochar or possessed reliable technology for producing it, and so had to restrict their biochar trials to smaller plot sizes. Others noted environmental issues as a concern, with one project respondent reporting that farmers were concerned about environmental safety (China) and another about the potential for deforestation (Africa).

This survey also investigated the prevalence of biochar as a traditional farming practice. Farmers are often conservative in their practices and it can be challenging to introduce new agricultural techniques, even when they are clearly beneficial. Intriguingly, in many regions, some form of biochar application was a traditional practice that was swept away by the advent of chemical fertilizers and other twentieth-century methods. The survey found that 35 percent of those who responded to this question reported a traditional indigenous use of biochar-like practices. Many felt that the existence of the traditional practice made their job of communicating the benefits of biochar much easier (appendix A).

In order to understand what drives the adoption of biochar systems, the final set of questions (appendix A) examined what the most important benefits were to project participants, including the potential benefit of carbon offset payments. As expected, soil improvement and crop yield increase topped the list, with many respondents specifically citing decreased fertilizer use and improved water use efficiency as important benefits. Clean cookstoves were next, followed by the hope of carbon payments. Projected income from selling biochar was somewhat important, and improved sanitation using biochar was also important to some projects. Several respondents also mentioned environmental hygiene from cleaning up waste in fields and in towns. Regarding the importance of carbon offset payments to project viability, respondents were split. Almost exactly half said that carbon payments would be nice but that they were not counting on them. Then, one remaining quarter said their project could not do without carbon payments and the other quarter replied they were not going to pursue them at all.

Note

1. E. Ferreira, personal communication, 2012.

CHAPTER 7

Potential Future Involvement of Development Institutions, Including the World Bank

Biochar: Knowns and Unknowns

This report has provided a current review of opportunities and risks of biochar systems, particularly in developing-country settings and for smallholders. By compiling a state of the art overview of current knowledge regarding biochar science, the report offers a reconciling view on different scientific opinions about biochar and biochar systems and offers an overall picture of the various advantageous perspectives of its science and application. In thoroughly reviewing the science around the soil and agricultural impacts of biochar application, the report shows the crucial importance of context, implying that no ready-made solutions exist as to what would constitute the ideal combination of feedstock waste, biochar pyrolysis unit, and biochar-to-soil application. However, through an in-depth survey of biochar systems currently in use around the world, the report contextualizes the current state of scientific knowledge on biochar. It also contributes new findings through the ISO-based life-cycle analyses of three different biochar systems in various contexts. In this chapter, some forward-looking suggestions are derived from the findings of previous chapters. The aim here is also to show how development institutions, including the World Bank, could support continued research and beneficial application of biochar in development.

Biochar and biochar systems, particularly on smaller scales and in developing-country settings, are a rather new proposition that might well prove to have positive impacts on the development–climate nexus under certain conditions. Potential areas of intervention could be manifold. Biochar systems can potentially link different sectors that are responsible for the promotion of green growth and development. Examples include situations where true waste streams can be successfully targeted and pyrolyzed with adequate technology, and the resulting biochar can then be applied to agricultural production systems. For example, a biochar system can offer a green waste management solution, which in addition leads to increased soil productivity, higher yields, and hence increased overall agricultural climate resilience. In fact, the life-cycle assessments conducted for this report show that a biochar system must include soil benefits to have greater emission reductions than bioenergy based on full combustion of biomass, and improved crop growth may provide in many cases greater financial returns to farmers than potential carbon markets. These gains in agricultural production in certain areas can be critical to reduce pressures on land. In this perspective, biochar can help address one of the most complex drivers of deforestation and improved use of biochar cookstoves could lead to lowered settlements of BC in the cryosphere.

However, as the previous chapters have shown, there are a number of uncertainties that need to be addressed in order to be able to proactively identify "no-regret" biochar interventions. In addition, it is important that decision tools are available to choose the appropriate biochar system technologies by recognizing the variability in design that will be required to respond to local environmental, agronomic, and social constraints and opportunities.

Further Research Needs

In general, the amount of funds that go into research and experimentation has not reached the level needed to scale up specific biochar systems comfortably. Life-cycle assessments (such as those described in this report) that indicate very practical directions for low-risk biochar opportunities are only beginning to appear in developing-country contexts, and in particular for small-scale systems. Issues regarding data quality and availability (especially time series data) for this kind of exercise nonetheless call for a fully monitored approach with replication of promising systems. Among the areas that deserve further research and assessment are the effective targeting of feedstocks that can be considered "true wastes," that is, they do not lead to a change in land use patterns with undesired leakage effects elsewhere, among other criteria. Furthermore, the development of pyrolysis units with technological specifications appropriate for developing countries is an area of further work. The biochar-to-soil application process, and the resulting effects on soil quality and yields, deserves further attention. A categorization of different biochars and their biochemical properties (resulting from different feedstocks and pyrolysis conditions) is a priority area for future work. This, in turn, would allow for better prediction of fertility effects, depending on the soil types and crops to which these biochars are applied. This knowledge will need to be made accessible to farmers either directly or via developing-country extension services and similar institutions.

The above clearly shows that further research is needed on many aspects of biochar systems, such as biochar characterization and soil health and productivity, but also on social aspects of biochar systems related to technology adoption. For instance, research is lacking on the increased workload that can come with certain biochar systems, which in turn could make them unattractive to farmers and women in particular. The life-cycle assessment in this report looked preliminarily at the impact on labor and suggests great variations between different biochar systems in terms of labor savings and increases. Another much-needed investigation would examine the viability of systems relying on small quantities of biochar. Proof-of-concept experiments to date have tended to use rather large quantities of biochar (in many cases over 10 tonnes per hectare), on small areas, to increase the likelihood of clear and publishable results. However, this does not always reflect well the actual limitations of smallholders in developing countries, who often have both limited available true waste feedstock for pyrolysis and limited production technology capacity, as is the case typically with stoves. Low application rate studies are needed to identify the impact of quantities of biochar as low as half a tonne per hectare or less on agricultural systems. Such studies would particularly benefit from a time series perspective.

The above suggests that it would be counterproductive to solely see these research needs as a *purely academic* effort. Scientific findings based on experiments conducted largely isolated from real-world developing-country conditions do only help to a limited extent in explaining many crucial aspects that relate to the introduction and ultimately successful adoption of a new technology such as biochar systems in a development context. There is a need to take a broader perspective in biochar studies, taking into account the trade-offs in the use of limited resources such as biomass and nutrients at the local level. While there may be accumulating evidence that specific biochar systems can deliver on the "triple win promise" (energy, climate, and food), all evidence indicates that no such specific biochar system can work as a silver bullet over large areas without considering local conditions.

Applied and long-term oriented research at scale of implementation thus seems to be an essential requirement. Here, institutions like the World Bank, particularly through its technical advisory and convening services, could help to forge effective alliances between the research community and development practitioners on the ground. A concrete example could include the establishment of an inventory of ongoing biochar studies in developing countries with a view to setting up long-term biochar study sites. One of the most critical gaps in knowledge regarding the value of modern biochars for agriculture and climate change mitigation precisely relates to the temporal brevity of most experiments. The Global Inventory of Long-Term Soil-Ecosystem Experiments, established by the Duke's Nicholas School of the Environment and Earth Sciences, is a good example of how such an applied scientific approach could work.[1] To date, only one developed-country biochar study is part of this network. The setting up and implementation of such a network across several different developing countries could be a concrete item for multilateral and bilateral donor support.

As this body of experiences expands, it will be possible to refine the criteria of desirable biochar interventions. The definition of these criteria could progressively lead to the establishment of biochar sustainability standards, which could then serve as a basis for policy regulation or certification schemes. Sustainability standards that take into account the multiple risks along the biochar chain (from biomass sourcing to production and soil application) are key to create consumer confidence, attract finance, and catalyze it toward sustainable biochar applications. Preliminary standards could be developed early, despite the gaps in knowledge, before an extensive period of testing throughout very different biochar experiences. Such standards are needed to provide guidance to project develop-ers on low-risk and viable biochar routes, also providing them with an alternative to thorough and costly life-cycle assessments.

Supporting the Early Adoption of Biochar Systems

To assist in the uptake of biochar projects, development institutions such as the World Bank could engage in knowledge- and technology-oriented services, as well as financing services for biochar projects or programs. Knowledge services could, for example, target the development of carbon finance-related method-ologies for different biochar systems, which would ultimately provide a public good allowing for different developing-country stakeholders to create carbon assets based on biochar systems. Fully capturing and monetizing the climate change mitigation-related potential of biochar systems may initially be impor-tant, given the challenge to adapt (some of the) smallholder-driven biochar sys-tems into a financially sustainable proposition. If, fundamentally, the viability of biochar systems for smallholders depends first and foremost on visible and sus-tained gains in soil quality and agricultural production, reaping potential carbon benefits could improve the overall economics of a given biochar system during periods of transition. At present, access to carbon markets is restricted, as was shown in the section titled "Engagement with Carbon Markets" in chapter 6. This is mainly due to the fact that no appropriate methodologies exist that would be able to quantify and convert the multiple climate change mitigation benefits of biochar systems along the entire value chain into a carbon asset.

Returning to the example of a cookstove-based biochar system, it would seem to be of interest to build on currently existing Clean Development Mechanism (CDM) as well as voluntary carbon market methodologies for cookstoves and to broaden their scope to include the specifics of cookstove biochar systems. Today, there are four methodologies approved under the CDM that projects and pro-grams targeting improved cookstoves and reduction of nonrenewable biomass can apply.[2] In addition, the Gold Standard has approved one voluntary carbon market methodology applicable to improved cookstoves (ClimateCare 2010). Both CDM and voluntary carbon markets are starting to include a growing num-ber of nonbiochar cookstove-related projects and programs. This shows that even without the added carbon value stemming from biochar, cookstove-based carbon finance operations and, particularly, programs do seem viable in certain develop-ing-country settings. A comprehensive review of these methodologies to identify elements that lend themselves to a biochar cookstove system might be a first concrete step toward the development of a more comprehensive biochar metho-dology or methodology modules. Such technical work could, for example, be undertaken through knowledge and capacity-building instruments attached to some of the World Bank's carbon funds. The BioCarbon Fund and the Community Development Carbon Fund have a grant-making capacity-building facility attached to them (called BioCF plus and CDCF plus, respectively). Work funded through those facilities could start with a comprehensive review of existing methodology elements and the development of new elements or modules along the value chain of a biochar system. The focus of such new methodology elements might not necessarily be the compliance-based carbon markets in the first place but, rather, the voluntary market, which shows a growing demand for agricultural and land-based carbon assets. After over 10 years of practice in the area of carbon finance it is evident that significant opportunities exist for rural populations to participate in the emerging global carbon market. Biochar application to soil can become one such option once methodologically fully covered by rewarding carbon management.

As mentioned before, the World Bank's carbon funds, administered by the Carbon Finance Unit of the World Bank and the International Finance Corporation, could finance the mitigation potential of biochar systems in parallel with the development of applicable methodologies. The rationale for engagement is a clear market failure to reward smallholder-based climate change mitigation benefits. In addition to climate- and carbon-related finance, the global environment facility (GEF) runs several grant mechanisms that could be used for promoting biochar cookstoves, for example, while improving the sustainability of household biomass use. These grants include the Earth Fund[3] (and other private sector development funds), the Sustainable Forest Management Program, and the Small Grants Program (GEF 2003, 2007, 2010). Within the recently approved GEF-5 envelope, one focal area identifies improved biomass cookstoves as a priority related to energy efficiency and sustainable forest management (GEF 2009). In addition, the GEF Small Grants Program, which supports the projects of nongovernmental organizations (NGOs) and community-based organizations, can also support cookstove projects (for a full review of potential cookstove-related funding that might apply to biochar cookstoves as well, see World Bank 2011).

While public resources are certainly needed for the early demonstration and research and development phases of biochar interventions, it will be crucial to also involve the private sector and stimulate opportunities for private investment early on. The step from publicly supported pilots to commercial viability is often the most difficult phase for many technologies and certainly also for biochar systems. Murphy and Edwards (2003) have called this critical transition period the "valley of death" and biochar system technology in a developing-country context will have to deal with this phenomenon. The valley of death is characterized by high investment costs and significant risks so that projects can easily fail. This is where neither technology-push nor market-pull forces have sufficient strength to bring a new technology to market viability. In this situation, neither the public nor the private sector might consider financing commercialization. This funding gap is particularly problematic for technologies with long lead times and a need for considerable applied research and testing between invention and commercialization, as is the case for many energy technologies (Norberg-Bohm 2002).

Therefore, innovative financing solutions such as, for example, trying to front-load potential carbon benefits of biochar systems might help to fill the financing gap in order to bridge the valley of death. Different instruments could be applied with the basic idea that the future stream of carbon emission reductions would be commercialized at the financial closure of the project. Thus, the up-front payment can contribute to the financing of the project even though the emission reductions are materialized and delivered at a later date once the project is commissioned and starts operation. In order to do such front-loading with the idea of attracting private investors early on, the emission reduction seller (that is, the entity operating the biochar system and implementing the project or program) would need to guarantee the investor a portion of the emission reduction delivery through a shortfall agreement, based on which the seller commits to pay for any emission reductions not delivered to the buyer at a pre-agreed price. Such a shortfall agreement is necessary to hedge against project and regulatory risks (that is, the stream of emission reductions does not materialize at all or not to the extent projected, or emission reductions generated are not eligible under an international trading scheme). These are still early ideas, but in principle, institutions such as the World Bank, particularly through their private sector-oriented branches, could possibly guarantee the aforementioned payment obligation of the seller based on an indemnity agreement between the seller and the Bank. It becomes clear that a certain level of sophistication is needed for such a setup and that institutions such as the World Bank would most likely not be able to offer such guarantee structures on a project-by-project basis but rather for a project aggregator or related entity working at larger scales (it could even be sovereign governments that bundle a program of biochar systems implemented for development purposes).

There are numerous ways in which biochar and biochar systems may provide "triple win" benefits in developing countries, and many ways that development institutions may help to facilitate and improve them. Direct financing of biochar projects is an important way in which the World Bank and other organizations can provide targeted and essential support to ensure sustainable development of biochar systems. As part of this process, a fund could be set up to convene open-source technology developers to accelerate technology development for smallholders, along the lines of the Global Alliance for Clean Cookstoves.

Scaling Up Biochar Systems from Pilot to Program

A deeper economic understanding of developing-country biochar systems is emerging. In many instances crop growth benefits for the rural household seem to outweigh the potential financial benefits of carbon markets, at least for the time being. Ultimately however, carbon mitigation-related financial revenues, which at present cannot be fully internalized due to the lack of applicable methodologies, may help to spur adoption rates. Biochar systems are a nascent technology and the private sector is not likely to easily engage on larger scales in developing-country situations due to some of the technical unknowns as well as the (soon to be predictable) variability described earlier. Therefore, institutions such as the World Bank and particularly the private sector arm of the World Bank, the International Finance Corporation, are key players in providing financing services for biochar projects in developing countries over the next few years. Publicly funded demonstration, research, and development will need the engagement of bilateral and multilateral development institutions. Initial guidance on a number of criteria to support decisions on the allocation of funding for biochar projects has been given in previous chapters of this report. In sum, the feasibility of biochar pilots may be rapidly assessed when considering key questions such as the following:

1. *Will the biomass be sourced from true waste sources?* Possible indicators here include that (a) in the absence of pyrolysis a net cost would arise associated with management of this waste biomass; (b) the waste would otherwise be combusted without energy capture; or (c) the waste would decompose in landfills without contribution to soil nutrients and carbon.
2. *Will the feedstock be sourced from safe materials?* Possible indicators should assess the toxicity of wastes, in particular from urban origins.
3. *Will the quantities of biochar required match the availability of suitable feedstock locally?* Possible indicators include (a) quantities of sustainable feedstock available on-farm; and (b) the potential to incorporate biochar in high-value cropping systems, such as vegetable gardens.
4. *Will the pyrolysis system meet certain levels of conversion efficiency and cleanliness?* Possible indicators include (a) methane emissions from the specific pyrolysis technology chosen; (b) a minimum threshold for carbon efficiency as a fraction of biomass input; and (c) emissions of particulates and toxic volatile products of incomplete combustion.
5. *Will the appropriate biochar be applied to appropriate soils?* Possible indicators here include (a) a modification in soil pH toward the optimum range (for example, 5.5–6.5 for most cereals); and (b) a higher water-holding capacity of biochar particles relative to the soil particles in environments limited by water.
6. *Will it be practical during monitoring activities to verify carbon storage of biochar through its application to soils?* Possible indicators can relate to (a) the variability of the biochar produced with regard to properties that determine its stability; and (b) difficulty of determining whether biochar is applied to soils rather than combusted for energy.
7. *Will local farmers likely adopt the technology after the demonstration phase?* Possible indicators include (a) whether access to land is constrained and whether gains in agricultural productivity is a social priority; (b) estimates of increased or reduced workload for women; and (c) generation of new and valuable energy services.

Apart from the fact that biochar systems are an area for public climate financing due to their mitigation potential, their major cobenefits do clearly qualify for additional funding sources related to themes such as energy access for the poor, improved rural health, and other environmental, agricultural, social, health, and economic benefits.

Facilitating the necessary applied and long-term research, and providing knowledge services such as carbon finance methodologies, are other key ways that organizations such as the World Bank could assist in building the knowledge base required by these nascent technologies. A feedstock survey to identify geographic regions with waste disposal problems could assist in indicating where initial deployment efforts can concentrate, serving as a roadmap for future development.

Research and development programs should be integrated with other initiatives, given the wide-ranging potential of biochar systems. For example, research on integrated food-energy systems (FAO 2010) could include biochar, building on the typology effort outlined in this report. Having a variety of production techniques available in a community could maximize the benefits of biochar to that community. Biochar-making stoves have many great advantages, but they may not produce enough biochar in the short term for large-scale row crops. There is potential for synergy through introducing biochar technology into communities at a range of scales, from stoves to make biochar for household uses (kitchen gardens, water filtration, sanitation), to larger units to make biochar commercially, with possibilities for electricity production or heat generation for crop drying or commercial cooking, brewing, and other activities.

Expanding on the theme of coordination with other efforts, ways should be found to integrate biochar into existing charcoal economies and clean charcoal efforts. Charcoal is in widespread use in much of the developing world as a fuel. In many cases the charcoal made as fuel would be perfectly suitable as biochar. Such an approach could help address the current unsustainability of charcoal production by using residues rather than forests, and using cleaner technology that is far more efficient than traditional methods. The Pro-Natura process is one example, the Adam retort kiln is another. Both of these were designed initially for better fuel charcoal, not biochar, but both are now being used to make biochar. From the standpoint of risk and investment, then, clean charcoal making is a low-risk investment, because even if the market for charcoal as biochar does not materialize, it is still a valuable product that can be sold as fuel or used for other purposes such as water filtration and sanitation. This helps avoid the "valley of death" in commercialization. From the standpoint of climate mitigation, clean charcoal is also a low-risk strategy. It is best if most of the charcoal produced gets used in agriculture, but even if it is used as fuel, if it substitutes for traditionally made charcoal, it has a positive climate benefit.

Finally, it is clear that detailed analyses of current projects need to be carried out across their life cycle in order to assess the costs and benefits, particularly with regard to climate, energy production, and food security. The LCA projects described in this report should be followed up over time, and new ones added. A project catalog should be maintained, and data collection should continue for all biochar projects. Lessons learned will be extremely valuable for future projects and programs, and systematic and robust systems should be set up for collecting data that can be of scientific value in demonstrating the implications of biochar systems for crops, climate, and human well-being.

Notes

1. Global Inventory of Long-Term Soil-Ecosystem Experiments: http://www.nicholas.duke.edu/ltse/.

2. (a) AMS I.C, version 18; (b) AMS I.E, version 4; (c) AMS II.G, version 4; and (d) AMS I.I, version 1.

3. Global Environment Facility website: The GEF Earth Fund http://www.thegef.org/gef/node/1293.

Survey Results

Part I

The survey had a number of open-ended questions and a space for general comments. This open-ended feedback was useful for identifying previously unknown applications and techniques. Open-ended questions also allowed for the capture of a list of crops grown with biochar and a list of feedstocks used to make biochar. A question about project goals highlighted that people are interested in biochar for many different purposes, including scientific research, development outcomes, climate impact, and commercial results. The results included here are those not shown directly in chapter 4. Figures A.1 to A.4 present some introductory data about the characteristics of the projects. The sample size of $n=149$ represents 149 complete survey responses.

Figure A.1 Project Phase ($n = 149$)

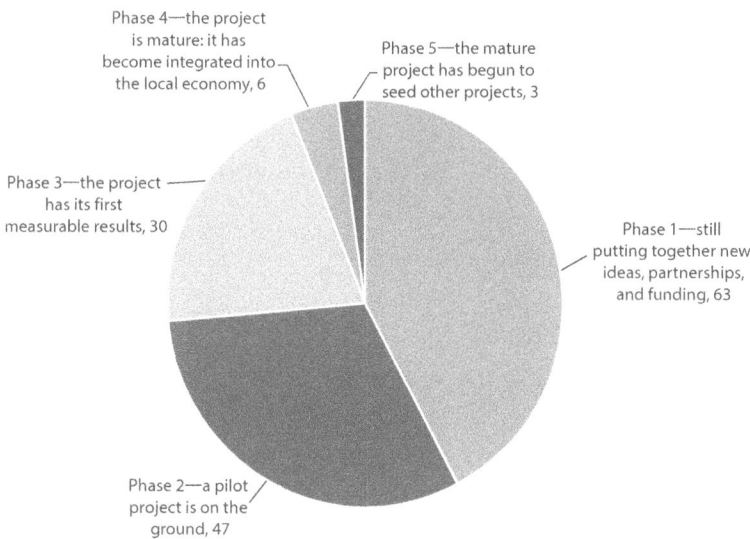

Source: World Bank.

Figure A.2 Project Location (*n* = 149)

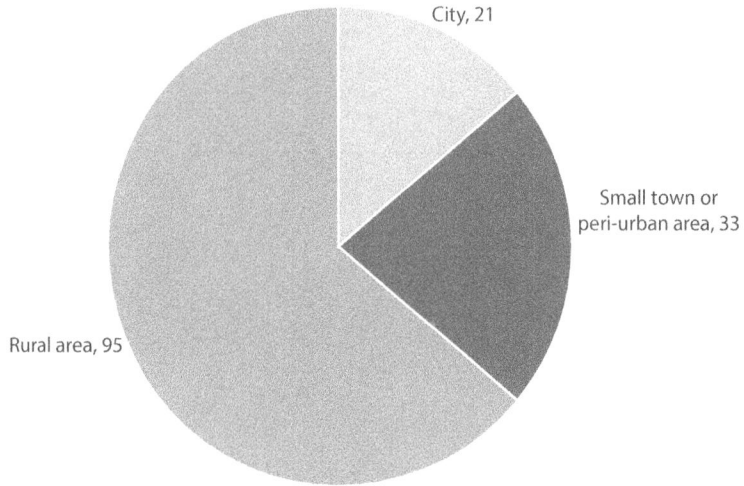

Source: World Bank.

Figure A.3 Utilization of Biochar Product (*n* = 442)

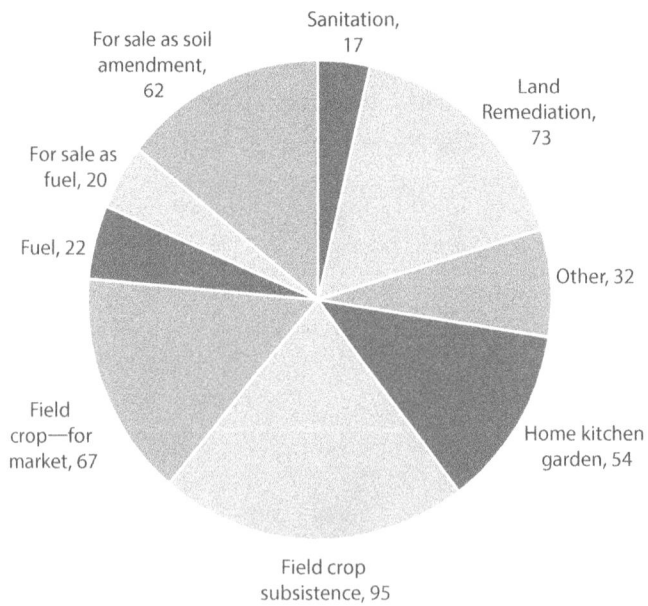

Source: World Bank.

Figure A.4 Biochar Pretreatments and Added Amendments (*n* = 359)

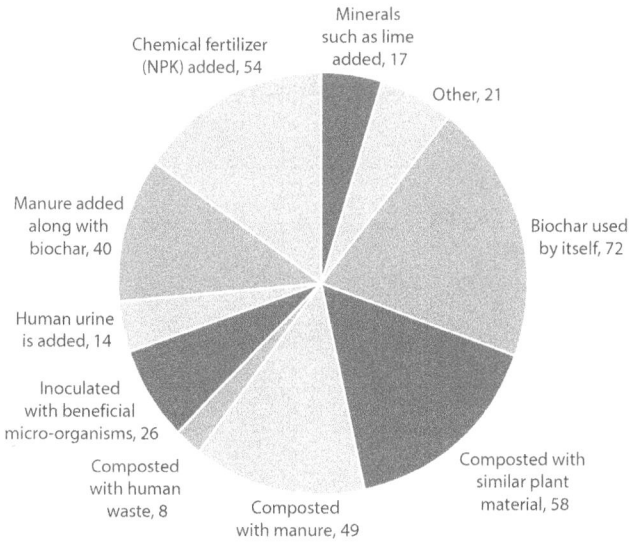

Source: World Bank.
Note: NPK = nitrogen, phosphorus, and potassium (fertilizer).

Part II

Figures A.5 to A.10, and tables A.1 to A.6, present data on factors surrounding projects, including social and cultural barriers and means of overcoming them; existence of biochar as a traditional practice; problems accessing biochar supplies; benefits foreseen from biochar projects; and the importance of carbon offset payments.

Figure A.5 Barriers to Implementing Biochar Systems (*n* = 39)

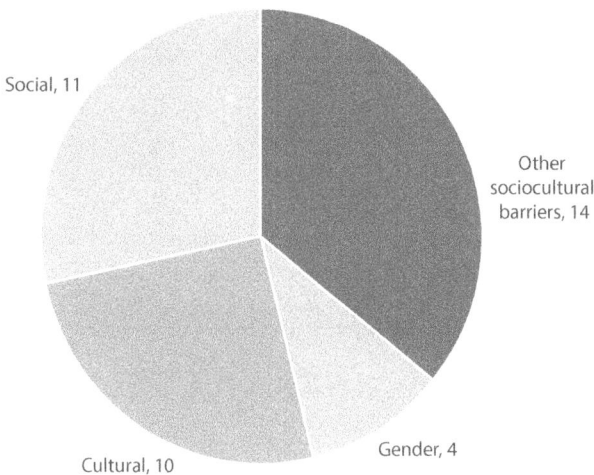

Source: World Bank.

Table A.1 Description of Barriers to Implementing Biochar Systems

Open-ended response to question:, "If you checked any of the items in the previous question, please describe the context and impact of the sociocultural barriers you encountered. Were the barriers related to biochar production or to application of biochar to soil?"

Description of biochar sociocultural barriers	Number of similar responses
People resist new approaches for development	3
Not enough biochar available to make a difference	3
Lack of technology for biochar production	3
Education, the farmers/authorities have no idea what biochar can do. Education programs are needed	3
Farmers are skeptical—they don't believe that the biochar is worth the cost and time	2
Relates to the application of biochar into the soil	1
Only young people are interested in biochar application	1
Language—much biochar information is in English only	1
Labor—slash and burn is more labor effective and gives short fast effect on crop yield	1
It might cause deforestation	1
Feedstock collection—it is difficult to collect feedstocks that are scattered and abundant all over the city in the form of spent coconut shells and husks, construction site leftovers, and garden wastes	1
Farmers are worried about its environmental safety	1

Source: World Bank.

Table A.2 Ideas for Overcoming Barriers

Open-ended response to question:, "Please describe any solutions you have thought of or tried in order to overcome social or cultural barriers to adoption of biochar systems."

Ideas for overcoming barriers	Number of similar responses
Lack of information—need to educate and demonstrate via small trials, farm visits, participatory research, agricultural extension	12
Technology access—development of indigenous technology	3
Cost-effectiveness—set up markets and carbon credits	2
Language barrier—working with local NGO and an English-fluent liaison officer	1
Money—get more money	1
Not enough biochar available for field crops—use it for vegetables	1
Biochar application—if funds could be made available for tractors to offer tilling of lands with biochar for selected farms at different locations within this farming district, others will follow suit	1

Source: World Bank.
Note: NGO = nongovernmental organization.

Figure A.6 Biochar as a Traditional Practice (*n* = 44)

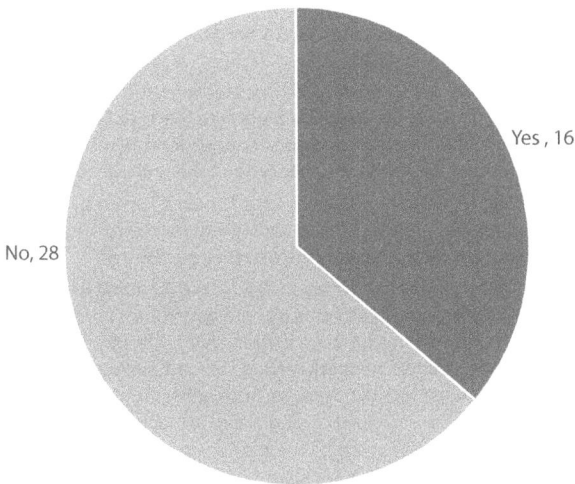

Yes, 16

No, 28

Source: World Bank.

Table A.3 Biochar as a Traditional Practice

All answers to the open-ended question:, "If you answered *yes* to the previous question, please describe the local tradition of adding black carbon char (not ash) to soil. How strong is this tradition? Is it being practiced today? Is the existence of this tradition helpful in facilitating the adoption of new biochar methods?"

"It is practiced by a few, although they don't know the other advantages of application. It has been helpful in some areas, and is now being adopted in other areas."

"This practice is strong in the Mekong delta region but not in central coastal Vietnam. In the Mekong area, there is industrial-scale use of rice husk biochar in agriculture, but not in central Vietnam (except for nurseries)."

"Historically added accidentally as part of slash-and-burn system, where larger branches became buried under hot ash and pyrolysis took place, forming charcoal—much of the response of new crops may have resulted from this."

"Within the agroforestry systems practices in the state of Karnataka, India, most farmers practice burning of "waste" materials, including prunings from tree crops, collected weed biomass, and other residues. This typically occurs directly on agroforestry fields. Although most of material is typically reduced to ash, there is likely to be creation of some amount of black carbon residuals that are incorporated into the field. Although not consciously or effectively practiced, black carbon incorporation is, to an extent, occurring."

"Sometimes the farmers get biochar through burning biomass in the field. They randomly apply the biochar in the soils."

"The farmers add the charcoal dust to soils for improving the status of soil fertility. They have experience of the coal production method and the impact of its dust in improving agricultural soils."

"It is a tradition known by the local elders but I have failed to locate any traditional owner doing it. However, I have created strong ties with the up-and-coming traditional owners who are traditional artists and they are getting me access to previously unreleased information."

"In India, older generations used to add cow dung cakes and ash to the soil along with char from wood to the fields to improve the carbon content of soil."

table continues next page

Table A.3 Biochar as a Traditional Practice *(continued)*

"Locally, it is a tradition here to plant ornamented trees in gardens and homes with what we call black soil, but not charcoal or biochar, and this knowledge makes it easier to explain biochar to the people, but the will to change overnight is the problem now, which needs complete orientation and education on radios and so on before we make headway."

"Mango farmers use black soil from an old dump site for nursery of their seedlings and so on, and this practice makes sense to them when NRO preaches the gospel of biochar to them, but what is left is massive adoption by all farmers. Our vegetable farmers and nurseries operators here rely solely on black soil from the old dump site most of the time, so this theory helps them understand the concept."

"Charcoal has been used by hobby gardeners and professionals, including in nurseries for growing orchids. Rural people also add black ash from traditional fire stoves to their garden soil."

"The ancestors of farmers in this area used to put the black carbon char in the land that was used for growing rice, especially in the area that lacked fertile soil. After the government promoted the use of chemicals 60 years ago, the tradition of using black carbon char for farming disappeared. However, at present, some farmers still see the black carbon char remaining in some areas of the farm that people in the old time period had used. Because of having this old tradition, we expect that the obstacles to the promotion of using the new biochar method will be less."

"Local people use the term *kono* for agricultural wastes that are piled and then burned; while the coals are lit, farmers spread the coals around the garden using a specific tool called a *konot*. Until today, the tradition still exists, and we hope that it will become the starting point for biochar production."

Source: World Bank.

Figure A.7 How Projects Cope with Limited Supplies of Biochar (*n* = 67)

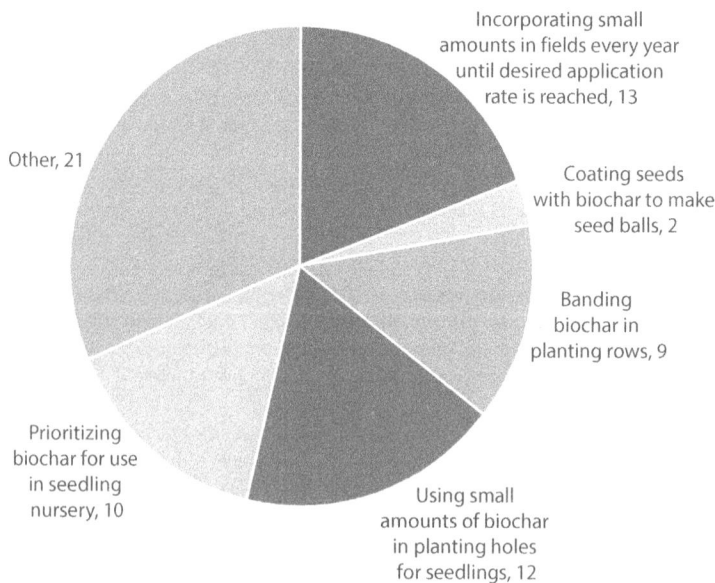

Source: World Bank.

Table A.4 Additional Ways to Cope with Limited Biochar Supplies

Responses to the open-ended question:, "If you like, please provide additional information about your biochar application rates and your ways of coping with limited supplies of biochar."

Other responses to limited biochar supplies	Number of similar responses
"Making more biochar—developing new production technology"	9
"Limiting our work to small experimental plots or nurseries"	5
"Using charcoal made for fuel use"	2
"We are banding around the canopy line of tree crops"	1
"Prioritizing use as follows: 1. vegetable plots; 2. rice fields; and 3. the area that we plant trees that can be used for producing energy (called *Jatropha curcas*)"	1
"Maximizing the multiple functions of biochar such as the bacteria carrier, using it as one of the components for fertilizer"	1

Source: World Bank.

Table A.5 Ranking of Benefits Project Developers Believe Will Accrue to Project Participants

Responses to the question, "How important are the following benefits to your project participants?"

	Not important		Somewhat important		Very important		Total	
	n	%	n	%	n	%	n	%
Soil improvement and soil conservation	1	2.3	5	11.4	38	86.4	44	100
Crop yield increase	2	4.3	3	6.5	41	89.1	46	100
Wages from production or processing of biochar	19	44.2	14	32.6	10	23.3	43	100
Income from sales of biochar	13	31.0	15	35.7	14	33.3	42	100
Improved sanitation using biochar	19	45.2	8	19.0	15	35.7	42	100
Water filtration using biochar	17	42.5	12	30.0	11	27.5	40	100
Cleaner cookstoves or kilns and improved air quality	17	40.5	8	19.0	17	40.5	42	100
Saving of labor or money to acquire fuelwood	18	46.2	10	25.6	11	28.2	39	100
Carbon offset payments for clean cookstoves or kilns	17	41.5	13	31.7	11	26.8	41	100
Carbon offset payments for use of biochar	9	22.0	17	41.5	15	36.6	41	100

Source: World Bank.

Table A.6 Further Description of Project Benefits

Responses to the open-ended question: "If you wish, please provide further descriptions of the benefits expected from the project."

Additional benefits from biochar systems	Number of similar responses
Increased sustainability by avoiding chemical fertilizers	7
Increased fertilizer use efficiency	6
Increased water use efficiency	5
Decreased N_2O and other GHG emissions	3
Environmental hygiene through waste reduction—fields	2
Environmental hygiene through waste reduction—cities	2
Improved pest resistance	1

Source: World Bank.
Note: GHG = greenhouse gas; N_2O = nitrous oxide.

Figure A.8 Responses to Ranking of Importance of Carbon Offset Payments to Project Finances ($n = 46$)

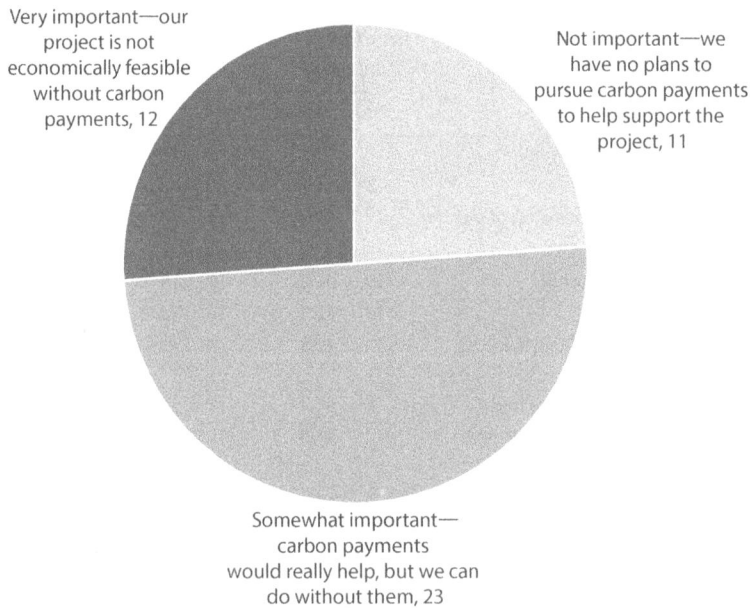

Very important—our project is not economically feasible without carbon payments, 12

Not important—we have no plans to pursue carbon payments to help support the project, 11

Somewhat important—carbon payments would really help, but we can do without them, 23

Source: World Bank.

Figure A.9 Responses to the Question of Who Would Receive Carbon Offset Payments from Biochar Projects (*n* = 35)

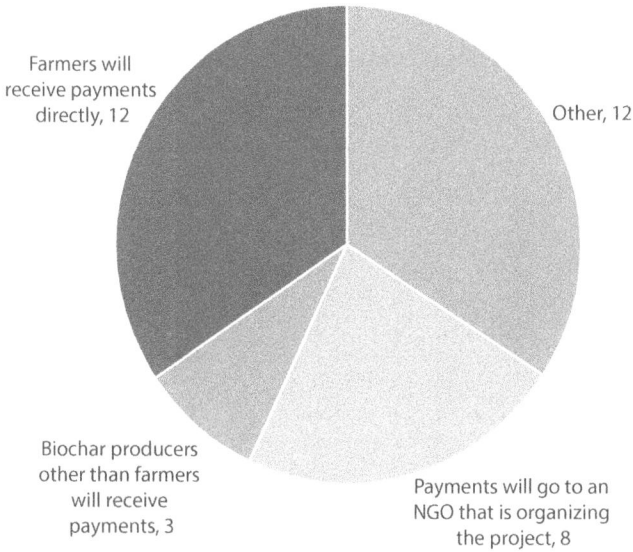

Source: World Bank.
Note: NGO = nongovernmental organization.

Survey Interpretation

Figures A.10, A.11, and A.12 present some analysis and interpretation of the survey results.

Figure A.10 Typology of Biochar Systems by Scale and Feedstock Showing Number of Projects with Each Type of Energy Recovery (*n* = 154)

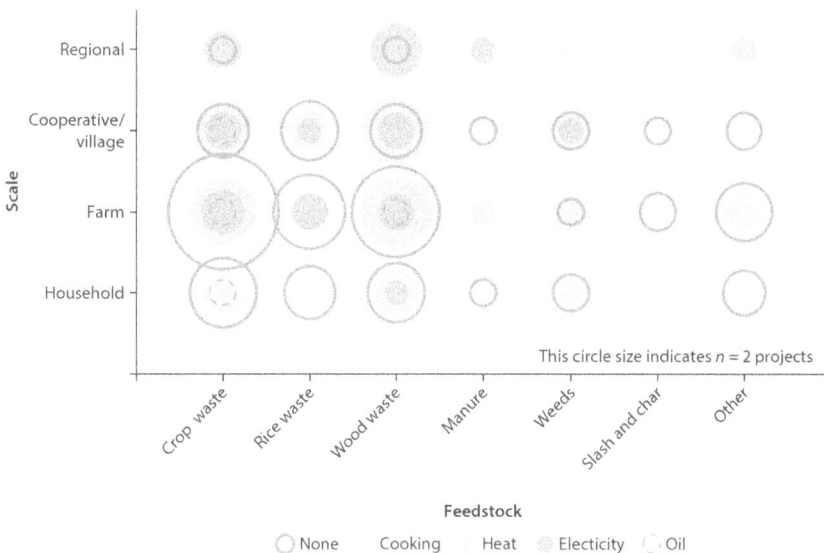

Source: World Bank.

Biochar Systems for Smallholders in Developing Countries • http://dx.doi.org/10.1596/978-0-8213-9525-7

Figure A.11 Typology of Biochar Systems by Energy Recovery and Feedstock Showing Number of Projects at Scale (*n* = 154)

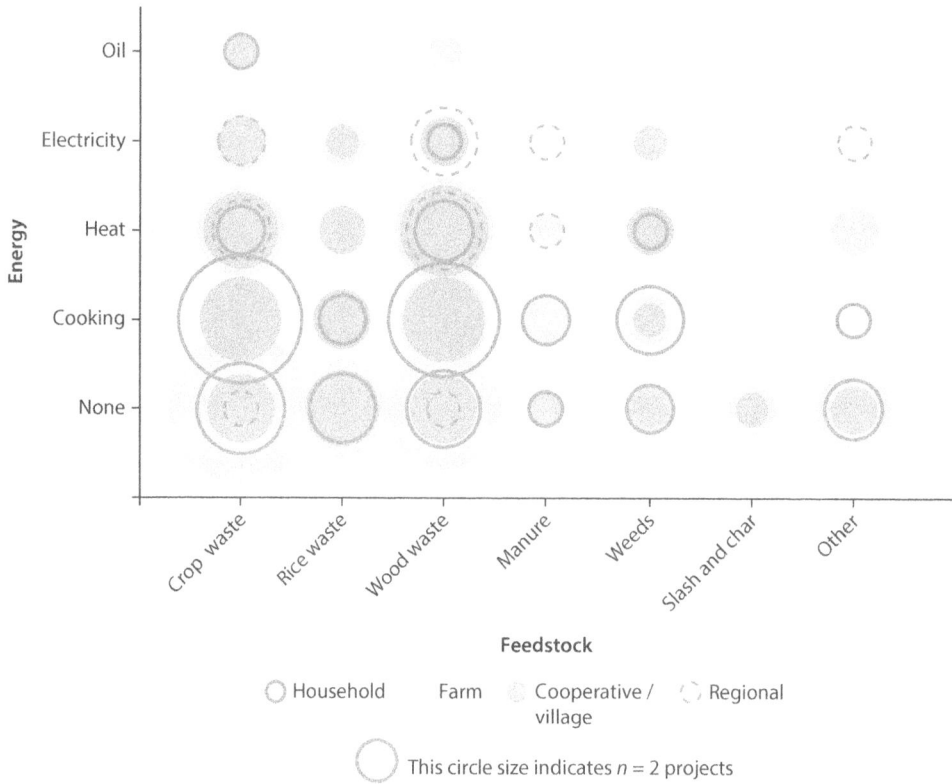

Source: World Bank.

A dendrogram is generated through the process of hierarchical clustering, whereby each point (in this case, each project) is grouped with similar points, and then these groups are grouped with similar groups, over and over again until there is only one large group, producing a figure that looks like an evolutionary tree. The closer the points are to each other, the more similar they are. The dendrogram was generated using the data on feedstocks, energy use, and scale for each project, and using the "centroid" clustering method.

Figure A.12 Project Dendrogram

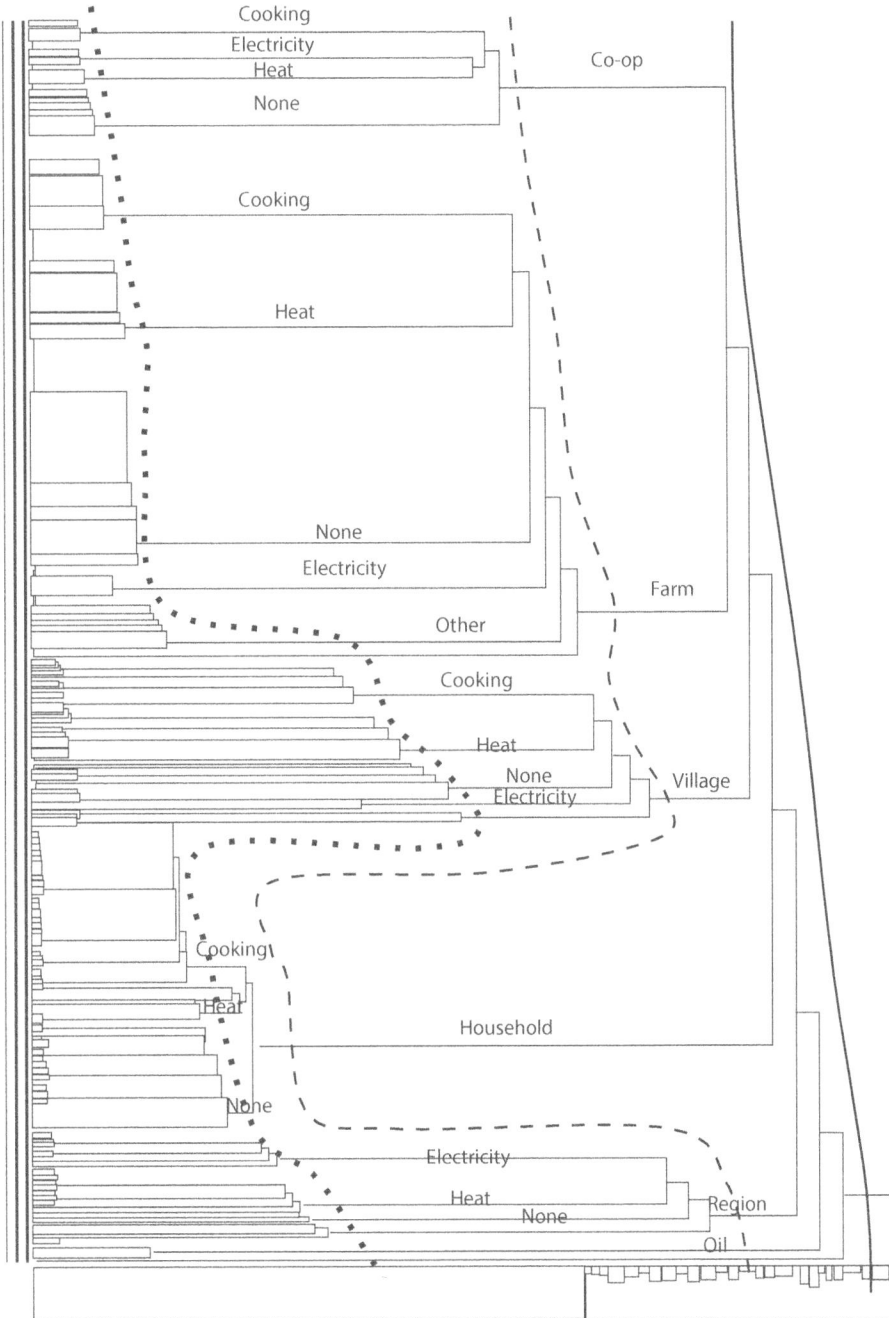

Source: World Bank.
Note: Solid line indicates the level at which branches contain a single project scale, dashed line indicates the level at which branches contain a single energy use, and dotted line indicates the level at which branches contain a single feedstock type.

Biochar Systems Survey: Project Data

Table A.7 Summary of Project Data from Survey

ID	Project location: country	Project goals	Scale of biochar production system	Biochar production technology	Energy use	Feedstocks used
91	Belize	To establish the efficacy of a biochar-based organic fertilizer on the cultivation of cacao and some food crops in Central America, looking specifically at soil fertility, yield data, soil resilience, and carbon sequestration	Cooperative	Batch retort kiln	No	Cacao prunings, shade tree prunings
118	Burkina Faso	The overall objective of the project is the promotion of renewable energies and environmental protection	Cooperative	Traditional pit, mound, or brick kiln	Cooking—household	Peanut and rice stalks
35	Cambodia	Carbon negative by poor nutrient soil amendment	Farm	Gasifier	Cooking—household, food, fuel or crop drying, electricity generation—internal combustion engine	Rice husk
158	Cambodia	The main project goal was to characterize the bioenergy/biochar system built around the gasifiers installed by SME Cambodia with respect to energy produced, biochar produced, carbon balance, life-cycle assessment of the gasifier, but also the rice product	Farm	Gasifier	Yes	Rice husk
132	Chile	Develop the technology for the production of biochar for agricultural use and forestry as an amendment to the ground and raw material base for the creation of an ecofertilizante-controlled release nitrogen compound	Farm	Pilot batch reactor	No	Oat husk, wheat husk, wheat straw, and pine sawdust
31	China	The objectives of our project were to investigate biochar addition on nutrient cycling (that is N and C) and greenhouse gas emissions from agroecosystems, mainly focusing on rice fields and vegetable ecosystems	Regional	Gasifier	Yes	
70	China	The function of biochar addition to soil	Village	Continuous kiln	No	
73	China	Developing biomass carbon engineering, agricultural application of char and related products	Regional	Continuous kiln	Yes	Wheat, corn, and animal waste
159	China	The goal of this project is to develop a policy suggestion on adopting biochar technology to reduce GHG emission to the Chinese government	Farm	Batch retort kiln	No	

ID	Project location: country	Project goals	Scale of biochar production system	Biochar production technology	Energy use	Feedstocks used
69	Costa Rica	Develop production and application standards for commercialization	Farm	Batch retort kiln	Food, fuel or crop drying, electricity generation—thermoelectric generator	Gmelina mill waste, bamboo, oil palm bunch waste
95	Costa Rica	Primary goal is to test and introduce biochar-producing clean cookstove to the rural agricultural workers' population in order to address health and deforestation issues; we will also incorporate a biochar demonstration garden in our project design	Household/family	Cookstove—top-lit up-draft gasifier (TLUD)	Cooking—household	Coffee prunings
166	Congo, Dem. Rep.	The Biochar Fund is in the process of conducting trials with farming communities in two selected regions: south-west Cameroon and the Equateur province of the Democratic Republic of Congo	Household/family	Batch retort kiln	No	Forest clearing slash, maize stover
42	Ghana	To prevent burning large quantities of coconut husk, improve soil fertility, and remove carbon dioxide to achieve 350 level	Farm	Traditional pit, mound, or brick kiln	No	Coconut husk, wood from cleared fields
141	Ghana	To use biochar from different feedstocks to improve soil productivity				Sawdust
22	India	To promote an innovative technology to produce and bury biochar to sequester carbon and improve soil fertility	Farm	Cookstove—top-lit up-draft gasifier (TLUD)	Cooking—household	Prosopis juliflora, crop residue, other biomass
125	India	To assess the effect of biochar produced from agricultural waste by ARTI's charcoaling technology, as a soil improvement agent in agriculture. To explore dose dependence of biochar addition on biomass yield. In general, to understand the barriers, if any	Cooperative	Batch retort kiln	No	Sugarcane trash, maize cobs
155	India	To study the long-term (at least five years) impact of biochar prepared from the residues of rice, wheat, corn, and pearl millet on productivity in rice-wheat, corn-wheat, and pearl millet-wheat cropping systems	Farm	Batch retort kiln	No	Rice, wheat, corn, and pea crop residues
13	Indonesia	Understand the impact of biochar application on growth and yield of crops and soils in lowland farming system on Aceh	Farm	Batch retort kiln	No	Rice husk, rice, straw

table continues next page

165

Table A.7 Summary of Project Data from Survey (continued)

ID	Project location: country	Project goals	Scale of biochar production system	Biochar production technology	Energy use	Feedstocks used
112	Indonesia	To get formulation of biochar-based soil conditioner with proper composition, shape, and dosage to improve soil quality	Farm	Batch retort kiln	No	Rice husk, oil palm shell
127	Indonesia	Evaluating potential and economic feasibility of rice husk and wood-derived biochars in West Java	Household/family	Traditional pit, mound, or brick kiln	No	Rice husk, small branches
36	Kenya	To improve and sustain our soil fertility for production. We also look forward to add human urine in the biochar compost to improve nitrogen in the soil to help in the control effects of striga infections	Household/family	Traditional pit, mound, or brick kiln		Acacia trees
58	Kenya	Remove water hyacinth from Lake Victoria and produce biochar as a by-product of cooking with water hyacinth briquettes	Farm	EveryThingNice stoves by WorldStove	Cooking—household, cooking—institutional	Water hyacinth from Lake Victoria
131	Kenya	This research study provides qualitative and quantitative data collected on the impact of adding charcoal to the soil. It provides preliminary data and information on the effect of charcoal application into the soil; particularly, in improving soil condition	Household/family	Traditional pit, mound, or brick kiln	Cooking—household	Stalks, sawdust, dry cow dung
144	Kenya	1. To develop a cheap gasifier stove for rural households. We have four years' experience with promotion of efficient clay stoves, and we performed successful experiments with metal gasifier stoves. 2. To find appropriate source of biomass for fuel	Household/family	Cookstove—top-lit up-draft gasifier (TLUD)	Cooking—household	Bagasse, wood
167	Kenya	Cookstove testing and feedstock assessment	Household/family	Cook stove—Anila, or annular retort design	Cooking—household	Maize stover
12	Malaysia	1. Production of biochar from rubberwood sawdust, coconut shells and EFB pellets in TLUD, Vesto, and traditional stove; 2. Physico-chemical characterization of biochar; 3. Pot assays (CRD) to identify the best biochar type	Household/family	Cookstove—top-lit up-draft gasifier (TLUD)	Cooking—household	EFB pellets, coconut shells
62	Malaysia	To characterize biochar derived from oil palm empty fruit bunch and investigate its effectiveness as a soil amendment for vegetables and field crops, and C sequestration	Farm	Commercial pyrolysis, pilot plant	Electricity generation—internal combustion engine	Oil palm empty fruit bunches (EFB)

ID	Project location: country	Project goals	Scale of biochar production system	Biochar production technology	Energy use	Feedstocks used
28	Mexico	Production, application, and evaluation of an inoculated biochar			Yes	Charcoal fines, mesquite and brush tree, sawdust
128	Nepal	Alternative fuel source in rural communities, income generation, reduce indoor air pollution	Village	Batch retort kiln	Cooking—household, space heating, food, fuel or crop drying, cooking—institutional	Catweed or hemp agrimony, forest wastes, agriculture wastes
105	Nigeria	Development of local capability for the production of activated carbons from agricultural wastes	Farm	Batch retort kiln	No	Coconut shells
79	Pakistan	To compare the effects of various integrated plant nutrient management practices on wheat-based cropping system under rain-fed conditions	Regional	Traditional pit, mound, or brick kiln	Food, fuel or crop drying	Crop residue, animal waste, and biopower
50	Philippines	Conversion of wood wastes and various agri-wastes into carbon (white charcoal) for household fuel, soil conditioner, water filters. To make people appreciate the value of charcoal in their daily lives	Cooperative	Batch retort kiln	Cooking—household	Tree bark, fruit frond, sawmill wastes
99	Philippines	To encourage farmers to organically produce farm products intended for human consumption	Farm	The biochar we used are waste products from rice hulls being used as fuel for mechanical dryers in our locality	Cooking—commercial, electricity generation—internal combustion engine	Rice husk
154	Senegal	Biochar production for local agronomic application	Cooperative	Pro-Natura pyrolyser Pyro	No	Rice husk
77	South Africa	Determine the effect of biochar application in South African and perennial orchard environments	Farm	Continuous kiln	Yes	Plantation waste
82	South Africa	To add carbon via fertilizer into soil	Regional	Pyrolysis	No	Wood chips and macadamia nut shells

table continues next page

Table A.7 Summary of Project Data from Survey *(continued)*

ID	Project location: country	Project goals	Scale of biochar production system	Biochar production technology	Energy use	Feedstocks used
117	South Africa	To manufacture biochar for use as a soil amendment in the aiREG bio-energy/agroforestry model	Cooperative	Continuous feed retort system	Food, fuel or crop drying, electricity generation—thermoelectric generator	Eucalyptus, pine, and alien vegetation
18	Vietnam	Finding out the adoptable biochar technology to improve the infertile soil crop yield with safety products and environment in Vietnam. Determining the good sources for BC production and develop in Vietnam	Household/family	Batch retort kiln	Cooking—household	Waste from maize, rice, wood
80	Vietnam	IBM			Food, fuel or crop drying	
87	Vietnam	Small-scale energy and biochar generation	Farm	Cookstove—top-lit up-draft gasifier (TLUD)	Cooking—household, food, fuel or crop drying	Rice hulls, coffee bean husks, wood shavings
140	Vietnam	Vietnam is an agricultural country with an agricultural land of over 9 M ha, of which there are light texture soils, consisting of 530 thousand ha of arenosols, and approximately 4.5 M ha of acrisols (National Institute for Soils and Fertilizers, 1996). These soils are poor	Household/family	Cookstove—top-lit up-draft gasifier (TLUD)	No	
160	Vietnam	To improve soil and water utilization in tree crop farming systems in central coastal Vietnam	Village	Gasifier	Cooking—institutional	Rice husk

Source: World Bank.
Note: GHG = greenhouse gas; BC = black carbon.

Case Studies

Kenya Case Study Life-Cycle Assessment

Data Quality Assessment

The survey had a number of open-ended questions and a space for general comments. This open-ended feedback was useful for identifying previously unknown applications and techniques.

The data quality for each process was assessed using the method from the University of Washington's Design for Environment Laboratory, based on the guidelines from Ansems and Ligthart (2002). Each process was assigned a data quality score for each indicator on a scale of 1 to 5, 1 being the best, and the average score was calculated for each process, as outlined in the goal and scope. The scoring of the processes is provided in table B.1. The majority of processes received an average score of 2. The process that received the poorest score was bus transport (2.4), because the data were from a diesel bus in the United States in 1998, as emissions data from buses in Kenya were not available.

Within the pyrolysis emissions subprocess, higher scores were received in the areas of geography, technology, representativeness, and reproducibility, due to the fact that the emissions data were not a direct measurement, but rather were calculated from ratios of emissions species for a gasifier cookstove. For the biochar application subprocess, higher scores were received in representativeness, reproducibility, and source, because the data were based on estimates for application methods. The cookstove production also received high scores in the areas of representativeness, reproducibility, and source, because the cookstoves had not been produced in quantity at that stage. Rather, the data were estimated based on the production of one prototype stove and the experiences of biomass cookstove researchers.

Table B.1 Data Quality Scores for Kenya Pyrolysis Cookstove System

Process	Time	Geography	Technology	Uncertainty	Representativeness	Reproducibility	Sources	Average score
Primary feedstock, on-farm, as collected	1	1	1	2	1	2	1	1
Primary feedstock, off-farm, as collected	1	1	1	2	1	2	1	1
Secondary feedstock, on-farm, as collected	1	1	1	2	1	2	1	1
Secondary feedstock, off-farm, as collected	1	1	1	2	1	2	1	1
Primary feedstock, on-farm, air-dry	1	1	2	2	2	3	2	2
Primary feedstock, off-farm, air-dry	1	1	2	2	2	3	2	2
Secondary feedstock, on-farm, air-dry	1	1	2	2	2	3	2	2
Secondary feedstock, off-farm, air-dry	1	1	2	2	2	3	2	2
Pyrolysis cooking	1	1	1	1	2	2	3	2
Pyrolysis emissions	1	3	3	2	3	3	1	2
Biochar production	1	1	1	2	2	2	3	2
Avoided traditional cooking	1	3	1	2	2	2	1	2
Avoided residue burning	1	3	2	2	2	2	1	2
Biochar, field application	1	1	1	2	3	3	3	2
Biochar, crop response	1	1	1	2	2	2	1	1
Cookstove production	1	1	1	2	3	3	4	2
Transport, bus	3	4	3	2	2	2	1	2

Source: World Bank.

Sensitivity Analysis

The parameters tested in the sensitivity analysis are listed in table B.2 (table 5.4 from chapter 5). In order to facilitate the data interpretation, the discussion of each sensitivity test is followed by a list of the results for those impact categories (greenhouse gas [GHG], money, and time) that are relevant to the test (that is, results are not presented if the value does not vary from the baseline).

The 1.15–3.02 tonnes of dry matter per household per year range for *primary* feedstock consumption is based on cooking fuelwood consumption data collected by two different methodologies, as described in chapter 5, subsection "Pyrolysis and Cooking." The results from the low (baseline) and high feedstock scenarios are shown in table B.3. The higher primary feedstock consumption results in a 61 percent higher net GHG reduction because of the larger amount of avoided emissions from cooking with a traditional three-stone stove. There is an additional 105 percent time savings with the higher feedstock consumption scenario, again because of the assumed higher fuelwood consumption for traditional cooking. (Note: the functional unit is 1 tonne of dry *secondary* feedstock, even though the parameter varied in this case is the *primary* feedstock.)

Table B.2 Sensitivity Analysis Input Parameters, Including Baseline and Range Values

Parameter	Baseline	Sensitivity range
Primary feedstock (tonnes of dry matter per household per year)	1.15	1.15–3.02
Stable carbon content of biochar (%)	80	0–90
Yield response with biochar additions (%)	+29	−50 to +97
Maize prices ($ per 90-kg bag)	0	18–36
Duration of biochar's effect (years)	50	1–100
Pyrolysis methane emissions (kg per tonne of dry matter)	1.98	1.64–6.4
Three-stone fire methane emissions (kg per tonne of dry matter)	5.17	0.60–6.4
Fraction of nonrenewable biomass for off-farm wood	0.8	0–1.0
Biochar application rate (tonnes per hectare)	27	2.7–27
Soil nitrous oxide emissions (%)	0	−50 to +50
Fraction of secondary feedstock collected off-farm	0	0–1.0

Source: World Bank.

Table B.3 Sensitivity Results for Primary Feedstock Parameter

Primary feedstock (tonnes of dry matter)	GHG (tonnes CO_2e per tonne of dry matter)	Change (%)	Time (hrs per tonne of dry matter)	Change (%)
1.15	−1.8	0	62	0
3.02	−2.9	+61	127	+105

Source: World Bank.
Note: GHG = greenhouse gas.

Table B.4 Sensitivity Results for Stable Carbon Content of Biochar Parameter

Stable carbon content of biochar (%)	GHG (tonnes CO_2e per tonne of dry matter)	Change (%)
0	−1.23	−31
50	−1.57	−11
80	−1.77	0
90	−1.83	+3

Source: World Bank.
Note: GHG = greenhouse gas; CO_2e = carbon dioxide equivalent.

The stable carbon content is varied in the range 0–90 percent with a baseline of 80 percent. A change in the stable fraction of carbon in the biochar from the baseline of 80 percent to 50 percent lowers the net GHG reductions by only 11 percent (table B.4). Meanwhile, a 90 percent stable carbon fraction increases the net GHG by 3 percent compared to the baseline. Even if the stable fraction of the carbon in the biochar is 0 percent, the net GHG is still −1.2 tonnes of carbon dioxide equivalent (CO_2e) per tonne of dry matter. From this analysis it is evident that the stable carbon content of the biochar is not a major driver in the net GHG balance of the Kenya household cookstove system.

Biochar Systems for Smallholders in Developing Countries • http://dx.doi.org/10.1596/978-0-8213-9525-7

The methane emissions during pyrolysis were varied in the range 1.64–6.4 kilograms of methane per tonne of dry fuel, based on low and high values found in the literature (table B.5). The low value of 1.64 kilograms of methane is based on low-power gasifier operation (MacCarty et al. 2008), while the high of 6.4 kilograms of methane is for an open fire (Johnson et al. 2008). For the lower emissions of 1.64 kilograms of methane per tonne of dry fuel, the net GHG reductions are increased by only 0.01 percent. On the higher emissions end at 6.4 kilograms of methane per tonne of dry fuel, the net GHG reductions decrease by 10 percent. Even if the pyrolysis cookstove emitted as much methane as an open fire, the net GHG balance for the biochar-producing household would only decrease by 10 percent. This indicates that the pyrolysis cookstove methane emissions do not dominate the net GHG emissions of the project life cycle within the uncertainty range.

The emissions range of 0.60–6.4 kilograms of methane per tonne of dry fuel for traditional three-stone fire cooking is also based on low and high values in the literature (table B.6). The 0.60 kilograms of methane is for a high-power three-stone fire (MacCarty et al. 2008), and the 6.4 kilograms of methane is from the open fire (Johnson et al. 2008). This range highlights the variability in emissions measurement results due to methodologies and experimental procedures, even for seemingly comparable technologies (the open fire and the three-stone fire). A higher methane emissions value for traditional cooking (6.4 kilograms of methane per tonne of dry fuel) results in higher avoided methane emissions, thus a 3 percent increase in net GHG reductions. For lower three-stone fire methane

Table B.5 Sensitivity Results for Pyrolysis Emissions Parameter

Pyrolysis methane emissions (kg of methane per tonne of fuel)	GHG (tonnes of CO_2e per tonne of dry matter)	Change (%)
1.64	−1.78	+0.01
1.98	−1.77	0
6.4	−1.59	−10

Source: World Bank.
Note: GHG = greenhouse gas; CO_2e = carbon dioxide equivalent.

Table B.6 Sensitivity Results for Three-Stone Fire Emissions Parameter

Three-stone fire methane emissions (kg per tonne of dry fuel)	GHG (tonnes CO_2e per tonne of dry mater)	Change (%)
0.60	−1.57	−11
5.17	−1.77	0
6.4	−1.82	+3

Source: World Bank.
Note: GHG = greenhouse gas; CO_2e = carbon dioxide equivalent.

emissions of 0.60 kilograms of methane per tonne of dry fuel, the net GHG reductions decrease by 11 percent. Similar to the pyrolysis cookstove emissions, the three-stone fire methane emissions do not dominate the GHG balance within the uncertainty range.

The duration of biochar's effect on crop productivity is varied from 1 year to 100 years, where the baseline is assumed to be 50 years. The "surplus maize" is the quantity of maize grown in excess of what would be grown without biochar additions over the 1-, 50-, or 100-year duration on the area to which biochar was applied by the household for that year (0.04 hectares). This assumes that the biochar is applied at the specified rate of 18 tonnes of carbon per hectare, rather than over the whole farm at a lower rate. Thus, the following year the family could gain additional benefits from the new area to which the biochar produced in that year was applied. If the $24 per bag of maize value is assigned to this surplus, the revenues are as listed in table B.7. It is important to note that these revenues are summed over the 1-, 50-, or 100-year duration of the biochar's effect on crop productivity. The net GHG balance changes only ±2 percent due to the effect of soil organic carbon accumulation over the duration of biochar's effect.

Maize prices are ranged from the baseline (no revenue for surplus maize) to up to $36 per 90-kilogram bag of grain, based on Kenya maize prices from 2000 to 2009 (Kirimi 2009). The "maize $" scenarios in figure 5.4 and discussed above utilize a $24 per bag price (based on August 2010 prices). Any monetary value for the maize results in an increased revenue for the household as compared to the baseline. However, one must consider that the baseline situation focuses on responsible household management and places the top priority on the family's nourishment. If maize grain was sold at the market but food in the home was insufficient, this is obviously a net loss for the family in terms of nutrition and health, regardless of maize sales. The sensitivity analysis compares varying the maize price for both a 1-crop effect and a 50-year effect for biochar on crop yields, where the 50-year effect exhibits strong sensitivity to the maize price. This comparison illustrates the importance of considering the sensitivity of another input parameter: the duration of biochar's agronomic effect. As the numbers in table B.8 demonstrate, even the lowest price for maize results in significant revenues when considering biochar's 50-year effect.

Table B.7 Sensitivity Results for Duration of Biochar's Effect on Productivity Parameter

Duration of biochar effect on productivity (yr.)	Surplus maize (kg per tonne of dry matter)	Revenue ($ per tonne of dry matter)	Change (%)	GHG (tonnes CO_2e per tonne of dry matter)
1	+15	+3	−99	−1.73
50	+739	+245	0	−1.77
100	+1,478	+491	+100	−1.80

Source: World Bank.
Note: GHG = greenhouse gas; CO_2e = carbon dioxide equivalent.

Biochar Systems for Smallholders in Developing Countries · http://dx.doi.org/10.1596/978-0-8213-9525-7

Table B.8 Sensitivity Results for Maize Price Parameter

Maize price ($ per 90-kg bag)	Revenue ($ per tonne of dry matter) 1-crop effect	Change (%)	Revenue ($ per tonne of dry matter) 50-year effect	Change (%)
0	−1.43	0	−1	0
18	+0.49	+135	+183	+184
24	+1.13	+179	+245	+246
36	+2.41	+269	+368	+369

Source: World Bank.

Table B.9 Sensitivity Results for Maize Yield Response Parameter

Yield response (%)	Surplus maize (kg per tonne of dry matter)	Change (%)	Revenue ($ per tonne of dry matter)	GHG (tonnes CO_2e per tonne of dry matter)
1-crop effect				
−50	−13	−263	−6	−1.72
0	0	−100	−1	−1.73
29	+8	0	+1	−1.73
97	+26	+225	+7	−1.73
50-year effect				
−50	−1,274	−272	−426	−1.66
0	0	−100	−1	−1.73
29	+739	0	+245	−1.77
97	+2,463	+233	+819	−1.86

Source: World Bank.
Note: GHG = greenhouse gas; CO_2e = carbon dioxide equivalent.

In order to consider the effect of varying the maize yield response with bio-char amendments, the surplus maize is valued at $24 per bag (even though the baseline assumes no monetary value for the surplus maize). A 1-crop biochar effect is also compared to a 50-year biochar effect in table B.9. The maize yield is ranged from –50 percent (a decrease in yield with biochar additions as compared to the control) to a 97 percent increase in yield (the maximum for fully fertilized maize plots from Kimetu et al. 2008). It is probable that smallholder farmers neither have access to the same quantities of fertilizers nor utilize the same farm management techniques as those in the Kimetu study. For this reason, it is possible that maize yields with biochar additions would vary within this range. When there is a 29 percent maize yield increase, there is an 8-kilogram surplus of grain per tonne of dry feedstock per crop, or 739 kilograms surplus grain per tonne of dry matter over 50 years. For the maximum yield increase of 97 percent, the net revenue is +$7 per tonne of dry matter and the surplus grain is 26 kilograms per tonne of dry matter for a 1-crop effect, while a 50-year effect results in $819 and 2,463 kilograms of grain per tonne of dry matter. With no change in crop yield, the net revenue actually goes negative and becomes a net cost at –$1 per tonne of dry matter. A 50 percent decrease in yield results in a more than 250 percent decrease in grain yield for both the 1-crop and 50-year biochar effects, and the net revenue is dominated by the lost maize sales. The changes in GHG emissions with a variation in maize yield are in the order of only a few kilograms of CO_2e per household per year, which is less than a fraction of a percent in change from the baseline for the 1-crop effect, and a 6 percent change or less for the 50-year effect. Small changes in net GHG are because the effect of yield increases on soil organic carbon is minimal, as discussed in the contribution analysis above. This also corresponds well with Whitman, Scholz, and Lehmann 2010, where the contribution of soil organic carbon to the overall GHG impact is only a fraction of the impact from avoided emissions and stable carbon in the biochar.

The fraction of nonrenewable biomass for off-farm wood is varied from 0 to 1, with a baseline of 0.8, as discussed in the life-cycle inventory data (table B.10). This analysis compares results for the 50-year biochar effect. The net GHG reductions decrease by 60 percent to –0.7 tonnes of CO_2e per tonne of dry matter when all of the biomass is assumed to be renewable. In this scenario, where the fraction of nonrenewable biomass (fNRB) equals 0, the avoided emissions from traditional cooking with a three-stone fire contribute only –0.2 tonnes of CO_2e per tonne of dry matter to the GHG reductions, as compared to the baseline, where avoided three-stone fire emissions are –1.3 tonnes of CO_2e per tonne of dry matter. When all of the off-farm wood is assumed to be nonrenewable (fNRB = 1), then the net GHG reductions increase by 15 percent to –2.0 tonnes of CO_2e per tonne of dry matter because all carbon dioxide emissions from avoided traditional cooking are included, as there would be no regrowth to uptake carbon dioxide emissions under this scenario.

The baseline biochar application rate is 18 tonnes of carbon per hectare, or 27 tonnes of biochar per hectare, where the biochar has an average carbon content of 66 percent. This application rate is relatively high and was used for research purposes, but it is possible that crop productivity improvements can be achieved at much lower application rates. Preliminary studies and unpublished data from Torres (2011) and Lehmann (2011) have seen increases in crop productivity with a biochar application rate of 2.6 tonnes per hectare, which is used as the lower bound for the sensitivity analysis using the 50-year biochar effect (table B.11).

Table B.10 Sensitivity Results for Fraction of Nonrenewable Biomass Parameter

fNRB	GHG (tonnes CO_2e per tonne of dry matter)	Change (%)
0	–0.71	–60
0.8	–1.77	0
1	–2.03	+15

Source: World Bank.
Note: fNRB = fraction of nonrenewable biomass; GHG = greenhouse gas; CO_2e = carbon dioxide equivalent.

Table B.11 Sensitivity Results for Biochar Application Rate Parameter

Biochar application rate (tonnes per hectare)	GHG (tonnes CO$_2$e per tonne of dry matter)	Change (%)	Surplus maize (kg per tonne of dry matter)	Change (%)	Revenue ($ per tonne of dry matter)
2.6	−2.12	+20	7,763	+950	+2,586
10	−1.83	+3	2,018	+173	+671
27	−1.77	0	739	0	+245

Source: World Bank.
Note: GHG = greenhouse gas; CO$_2$e = carbon dioxide equivalent.

Table B.12 Sensitivity Results for Soil Nitrous Oxide Emissions Parameter

Soil nitrous oxide emissions (% of baseline)	GHG (tonnes CO$_2$e per tonne of dry matter) 1-crop effect	Change (%)	GHG (tonnes CO$_2$e per tonne of dry matter) 50-year effect	Change (%)
−50	−1.74	+0.01	−2.54	+44
0	−1.73	0	−1.77	0
+50	−1.72	−0.01	−1.00	−44

Source: World Bank.
Note: GHG = greenhouse gas; CO$_2$e = carbon dioxide equivalent.

At the 2.6 tonnes per hectare application rate, the net GHG reductions increase by 20 percent, while the surplus maize increases an order of magnitude to 7,763 kilograms per tonne of dry matter. Or, the 10 tonnes per hectare application rate (which is a common application rate for biochar projects) results in a 3 percent increase in GHG reductions while the surplus maize increases to 2,018 kilograms per tonne of dry matter. Varying the biochar application rate has a very strong impact on the quantity of surplus maize produced per tonne of feedstock, that is, the same amount of feedstock is converted to biochar but can make a larger impact when the application rate is lower and the biochar can be spread over a larger area (assuming the agronomic effect is the same).

Soil nitrous oxide emissions are varied from +50 percent to −50 percent, estimated from literature values (Singh et al. 2010). This analysis compares the 1-crop and the 50-year biochar effects (table B.12). The changes in net GHG reductions are only ±0.01 percent when considered on a 1-crop basis. On a 50-year biochar effect basis, the soil nitrous oxide emissions could play a more substantial role at ±44 percent of the baseline. This analysis emphasizes the need for improved data on biochar's role in soil nitrous oxide emissions.

The fraction of secondary feedstock (such as crop residues and other herbaceous sources) sourced on-farm is ranged from 0 to 1.0, where the baseline is 0 (meaning all secondary feedstock required for the pyrolysis cookstove is supplied on the farm). This scenario is considered because of the potential need to retain all crop residues on the fields for erosion protection, thus requiring off-farm collection from sources such as invasive weeds and sawmill wastes. The life-cycle assessment (LCA) assumes that all secondary feedstock collected both on the farm and off the farm is from renewable sources. If residues are collected off-farm, the soil organic carbon depletion from the on-farm residue harvest is eliminated, but the net impact on the GHG balance is small, as this is only a small contribution to the total (4 kilograms of CO_2e per tonne of dry matter), as seen in figure 5.3. However, the impact on the net labor of the system would be substantial because of the time spent collecting secondary feedstock off the farm. Although there are no data available specific to how far and what types of herbaceous residues could be collected off the farm, if the time to collect the residues is assumed to be the same as for off-farm wood (and assuming a moisture content of 50 percent for herbaceous biomass), then the net labor changes to −65 hours per tonne of dry matter, meaning that even though there are labor savings from reduced off-farm primary fuel collection (as in the baseline scenario), time is now lost due to increased off-farm secondary feedstock collection. Alternatively, if one assumes that secondary feedstock could be collected closer to home and the time to collect is halved, the net labor of the system is −2 hours per tonne of dry matter, thus still a net loss in labor. However, one benefit that would remain under this scenario is that the off-farm harvest of nonrenewable wood for traditional cooking could be replaced by off-farm harvest of renewable herbaceous feedstocks.

Vietnam Case Study Life-Cycle Assessment

Data Quality Assessment

Each process was assigned a data quality score for each indicator on a scale of 1 to 5, 1 being the best. The average of the scores was calculated for each process, as provided in table B.13. The majority of processes received an average score of 2. The process that received the poorest score was the rice wafer stove construction (2.7) because the stove construction and operation are outside the scope of the biochar project. The rice wafers are produced regardless of the biochar project, and have been for some time, thus it has not been a high priority for the project participants. The next highest score was for the biochar wetting process, as this again is part of the stove operation, and no experimental data have been measured. The moisture content has been estimated, and from this the approximate volume of water needed to wet the biochar.

The stove construction, biochar production, and biochar wetting all received scores of 4 in the "uncertainty" category because the mean value, standard deviation, uncertainty type, and description of strengths and weaknesses were not available and could be approximated. The nitrogen, phosphorus, and potassium (NPK) fertilizer productions each received scores of 4 in geography because the data are specific to the United States. The data quality assessment highlights that there are concerns over uncertainty and representativeness in the data, and weaknesses in the data are explored in the sensitivity analysis.

Table B.13 Data Quality Scores for Vietnam Rice Husk Biochar System

Process	Time	Geography	Technology	Uncertainty	Representativeness	Reproducibility	Sources	Average score
Rice husk, at mill	1	2	1	2	3	2	3	2
Stove production	4	1	1	4	3	3	3	3
Biochar production	1	1	1	4	3	3	2	2
Biochar, wetting	1	1	1	4	3	4	3	2
Biochar, field application	1	1	1	2	2	2	2	2
Manure production	1	1	1	3	3	3	3	2
Transport, bullock cart	1	1	1	3	3	3	3	2
Nitrogen fertilizer production	1	4	2	2	3	1	1	2
P_2O_5 fertilizer production	1	4	2	2	3	1	1	2
K_2O fertilizer production	1	4	2	2	3	1	1	2
Avoided residue burning	1	3	1	2	2	2	1	2

Source: World Bank.
Note: P_2O_5 = phosphorus pentoxide; K_2O = potassium oxide.

Table B.14   Sensitivity Analysis Input Parameters, Including Baseline and Range Values

Parameter	Baseline	Sensitivity range
Biomass throughput (tonnes of dry matter per year)	26	
Stable carbon content of biochar (%)	80	0–90
Yield response with biochar additions (%)	+16	−50 to +47
Peanut prices ($ per tonne)	750	682–818
Rice wafer stove methane emissions (kg per tonne of dry matter)	2.24	1.64–6.4
Avoided rice husk burning methane emissions (kg per tonne of dry matter)	3.71	3.71–25.7
Duration of biochar's effect (years)	50	1–100
Biochar transportation distance (km)	5	1–25
Soil nitrous oxide emissions (%)	0	−50 to +50

Source: World Bank.

Sensitivity Analysis

A sensitivity analysis was conducted in order to measure the variability in the LCA results as a function of varying key input parameters. The input parameters that were tested are listed in table B.14 (table 5.12 from chapter 5), and include the fraction of recalcitrant carbon in the biochar, the yield response of peanut crops with biochar additions, the price the farmer receives for peanuts, methane emissions of the rice wafer stove, methane emissions from avoided rice husk burning, the duration of biochar's agronomic effectiveness, the biochar transportation distance, and soil nitrous oxide emissions. In order to facilitate the data interpretation, the discussion of each sensitivity test is followed by a list of the results for those impact categories (GHG or monetary value) that are relevant to the test (that is, results are not presented if the value does not vary from the baseline).

The stable carbon content is varied in the range 0–90 percent with a baseline of 80 percent (table B.15). A change in the stable fraction of carbon in the biochar from the baseline of 80 percent to 50 percent lowers the net GHG reductions by 29 percent. Meanwhile, a 90 percent stable carbon fraction increases the net GHG by 10 percent compared to the baseline. Even if the stable fraction of the carbon in the biochar is 0 percent, the net GHG is still −0.1 tonnes of CO_2e per tonne of dry matter. From this analysis it is evident that the stable carbon content of the biochar is a major driver in the net GHG balance of the Vietnam rice husk biochar system.

The methane emissions during rice wafer cooking and biochar production were varied from 1.64 to 6.4 kilograms of methane per tonne of dry fuel, based on low and high values found in the literature, where the baseline value is 2.24 kilograms per tonne of dry fuel (table B.16). The low value of 1.64 kilograms of methane is based on low-power gasifier operation (MacCarty et al. 2008), while the high of 6.4 kilograms of methane is for an open fire (Johnson et al. 2008). For both the lower and upper bounds of the emissions, the net GHG reductions change by less than 0.01 percent. This indicates that the pyrolysis cookstove methane emissions do not dominate the net GHG emissions of the project life cycle within the uncertainty range. The allocation assigned to the biochar production affects this impact, where if 100 percent of the impacts of rice wafer cooking were assigned to biochar production, then the result changes two orders of magnitude more than at only 0.6 percent allocation, but the net GHG still only change by less than 1 percent.

Table B.15 Sensitivity Results for Stable Carbon Content of Biochar Parameter

Stable carbon content of biochar (%)	GHG (tonnes CO_2e per tonne of dry matter)	Change (%)
0	−0.1	−77
50	−0.4	−29
80	−0.5	0
90	−0.6	+10

Source: World Bank.
Note: GHG = greenhouse gas; CO_2e = carbon dioxide equivalent.

Table B.16 Sensitivity Results for Rice Wafer Stove Methane Emissions Parameter

Rice wafer stove methane emissions (kg of methane per tonne of fuel)	GHG (tonnes CO_2e per tonne of dry matter)	Change (%)
1.64	−0.51237	< 0.001
2.24	−0.51239	0
6.4	−0.51231	< −0.001

Source: World Bank.
Note: GHG = greenhouse gas; CO_2e = carbon dioxide equivalent.

The emissions range of 3.71–25.7 kilograms of methane per tonne of dry fuel for rice husk burning is also based on low and high values in the literature (table B.17). The low value of 3.71 kilograms of methane per tonne of dry fuel is as described in the process data previously (for rice husk burning in smoldering piles). The high value of 25.7 kilograms of methane per tonne of dry fuel is from rice husk piles in a wind tunnel simulating open-air burning (Yonemura and Kawashima 2007). This range highlights the variability in emissions measurement results due to methodologies and experimental procedures, feedstock material, and environmental factors. A higher methane emissions value for avoided burning results in higher avoided methane emissions, thus a 0.8 percent increase in net GHG reductions. Similar to the rice wafer stove emissions, the avoided rice husk methane emissions do not dominate the GHG balance within the uncertainty range.

The duration of biochar's effect on crop productivity is varied from 1 year to 100 years, where the baseline is assumed to be 50 years (table B.18). The "surplus peanuts" are the quantity of peanuts grown in excess of what would be grown without biochar additions over the 1-, 50-, or 100-year duration on the area to which biochar was applied by the farmer. This assumes that the biochar is applied at the specified rate of 20 wet tonnes per hectare. The $750 per tonne value is assigned to the peanut surplus. It is important to note that these revenues are summed over the duration of biochar's effect on crop productivity (from as short as 1 year to as long as 100 years). Note also that there are two cropping seasons per year. The net GHG balance varies 20–30 percent, primarily due to the avoided phosphorus and potassium fertilizers and also to the effect of soil organic carbon accumulation to a smaller extent.

Peanut prices are ranged from $682 to $818 per tonne, with a baseline of $750 per tonne, based on Vietnam peanut prices in that region.[1] The sensitivity analysis compares varying the peanut price for the 50-year effect for biochar on crop

Table B.17 Sensitivity Results for Rice Husk Burning Methane Emissions Parameter

Rice husk burning methane emissions (kg methane per tonne of fuel)	GHG (tonnes CO_2e per tonne of dry matter)	Change (%)
3.71	−0.512	0
25.7	−0.516	+0.8

Source: World Bank.
Note: GHG = greenhouse gas; CO_2e = carbon dioxide equivalent.

Table B.18 Sensitivity Results for Duration of Biochar's Agronomic Effect Parameter

Duration of biochar's effect on productivity (yr)	Surplus peanut (kg per tonne of dry matter)	Revenue ($ per tonne of dry matter)	Change (%)	GHG (tonnes CO_2e per tonne of dry matter)
1	+16	+15	−98	−0.4
50	+822	+948	0	−0.5
100	+1,644	+1,900	+100	−0.6

Source: World Bank.
Note: GHG = greenhouse gas; CO_2e = carbon dioxide equivalent.

yields. As the numbers in table B.19 demonstrate, even the lowest price for peanuts results in significant revenues when considering biochar's 50-year effect,
where the change in the net economic balance is relatively small, at ±6 percent.

In order to consider the effect of varying the crop yield response with biochar
amendments, a 1-crop biochar effect is compared to a 50-year biochar effect in
table B.20. The peanut yield is ranged from −50% (a decrease in yield with biochar
additions as compared to the control) to a 47 percent increase in yield (the maximum percent increase for peanut plots receiving biochar alone compared to no
amendments) (Slavich et al. 2010).[2] At the baseline value of 16 percent, there is
an 8-kilogram surplus of peanuts per tonne of dry feedstock per crop, or an
822-kilogram surplus per tonne of dry matter over 50 years. For the maximum
yield increase of 47 percent, the net revenue is increased to +$19 per tonne of dry
matter and a surplus of 24 kilograms of peanuts per tonne of dry matter for a
1-crop effect, while a 50-year effect results in $2,158 and 2,423 kilograms of pea-
nuts per tonne of dry matter. With no change in crop yield, the net revenue
decreases but remains positive for both the 1-crop and 50-year effects, as the offset
manure and fertilizer still provide savings. A 50 percent decrease in yield results in
a more than 400 percent decrease in peanut yield for both the 1-crop and 50-year
biochar effects, and the net revenue is dominated by the lost peanut sales. The
changes in GHG emissions with a variation in crop yield are less than a fraction of
a percent from the baseline because the effect of yield increases on soil organic
carbon is minimal, as discussed in other case studies.

Table B.19 Sensitivity Results for Peanut Price Parameter

Peanut price ($ per tonne)	Revenue ($ per tonne of dry matter)	Change (%)
682	+892	−6
750	+948	0
818	+1,004	+6

Source: World Bank.

Table B.20 Sensitivity Results for Yield Response Parameter

Yield response (%)	Surplus peanuts (kg per tonne of dry matter)	Change (%)	Revenue ($ per tonne of dry matter)	Change (%)
1-crop effect				
−50	−26	−425	−19	−400
0	0	−100	+1	−86
16	+8	0	+7	0
47	+24	+200	+19	+157
50-year effect				
−50	−2,590	−415	−1,612	−270
0	0	−100	+331	−65
16	+822	0	+948	0
47	+2,423	+195	+2,158	+128

Source: World Bank.

Biochar Systems for Smallholders in Developing Countries • http://dx.doi.org/10.1596/978-0-8213-9525-7

The transportation distance of the biochar is ranged from 1 to 25 kilometers, with a baseline of 5 kilometers (table B.21). Even at a distance of 25 kilometers, the net revenue decreases by only 3 percent for the 50-year biochar effect, corresponding to the increased transportation cost of $2.05 per tonne-kilometer of biochar. However, considering a 1-crop effect, a 25-kilometer distance makes the yearly net revenue negative at −$17 per tonne of dry matter. In fact, a biochar transportation distance greater than 10 kilometers will cause the yearly net revenue to go from a positive to negative balance for the 1-crop effect. Therefore, even with a 50-year biochar effect, if transportation distances for the biochar are farther than 10 kilometers, it is possible that the first few peanut seasons may see a net loss, which is balanced by increased revenues in later years.

Soil nitrous oxide emissions are ranged from +50 percent to −50 percent, estimated from literature values (Singh et al. 2010) (table B.22). This analysis compares the 1-crop and the 50-year biochar effects. The changes in net GHG reductions are ±1 percent when considered on a 1-crop basis, demonstrating that biochar's effect on soil nitrous oxide emissions are minimal when considered only in the short term. However, for the cumulative 50-year biochar effect, the soil nitrous oxide emissions could play a more substantial role at more than ±100 percent of the net GHG of the baseline. The change in soil nitrous oxide emissions as a result of biochar's interaction with nitrogen fertilizer applications is summed over the 50 years with two crops per year, and becomes a very important parameter at this scale. These results, along with the Kenya cookstove case study, emphasize the need for improved data on biochar's role in soil nitrous oxide emissions.

Table B.21 Sensitivity Results for Biochar Transportation Distance Parameter

Biochar transport distance (km)	Revenue ($ per tonne of dry matter)	Change (%)
1	953	+0.5
5	948	0
25	924	−3

Source: World Bank.

Table B.22 Sensitivity Results for Soil Nitrous Oxide Emissions Parameter

Soil nitrous oxide emissions (% of baseline)	GHG (tonnes CO_2e per tonne of dry matter) 1-crop effect	Change (%)	GHG (tonnes CO_2e per tonne of dry matter) 50-year effect	Change (%)
−50	−0.420	+1	−1.1	+102
0	−0.415	0	−0.5	0
50	−0.409	−1	+0.03	−106

Source: World Bank.
Note: GHG = greenhouse gas; CO_2e = carbon dioxide equivalent.

Senegal Case Study Life-Cycle Assessment

Data Quality Assessment

Each process was assigned a data quality score for each indicator on a scale of 1 to 5, 1 being the best. The average of the scores was calculated for each process, as provided in table B.23. Each process received an average score of 2. The process that received the poorest score was the pyrolysis unit construction (2.4), because the construction materials were based on a rough estimation from the project contacts. The next highest score was 2.3 and was received by the fossil fuel production and combustion processes (residual oil, truck transport, and so on) as these are based on U.S. data.

The Pyro-6F construction and biochar production received scores of 4 in the "uncertainty" category because the mean value, standard deviation, uncertainty type, and description of strengths and weaknesses were not available and could not be approximated. The fossil fuel production and combustion processes each received scores of 4 in geography because the data are specific to the United States. The data quality assessment highlights that there are concerns over uncertainty and representativeness in the data, and weaknesses in the data are explored in the sensitivity analysis.

Sensitivity Analysis

A sensitivity analysis was conducted in order to measure the variability in the LCA results as a function of varying key input parameters (table B.24). The input

Table B.23 Data Quality Scores for Senegal Biochar System

Process	Time	Geography	Technology	Uncertainty	Representativeness	Reproducibility	Sources	Average score
Rice husk, at mill	1	1	2	2	2	2	1	2
Rice husk, bagged	1	1	2	2	2	2	1	2
Pyro unit construction	1	1	1	4	3	4	3	2
Biochar production	1	1	1	4	3	3	3	2
Residual oil, utility boiler combustion	1	4	3	2	2	2	2	2
Biochar, field application	1	1	1	2	2	2	3	2
U.S. class 8B diesel truck	1	4	3	2	2	2	2	2
U.S. barge: residual fuel	1	4	3	2	2	2	2	2
Transport, pyro unit	1	1	3	2	2	2	2	2
Transport, bullock cart	1	1	1	3	3	3	3	2
Residual oil, at point of use	1	4	3	2	2	2	2	2
Electricity	1	2	3	2	3	3	2	2
Avoided rice husk decay	1	2	2	2	3	3	2	2
Average steel, 30% virgin, 70% recycled	1	4	3	2	2	2	2	2

Source: World Bank.

Table B.24 Sensitivity Analysis Input Parameters, Including Baseline and Range Values

Parameter	Baseline	Sensitivity range
Biomass throughput (tonnes of dry matter per year)	1,388	
Stable carbon content of biochar (%)	80	0–90
Yield response with biochar additions (%)	+52	−50 to +233
Onion price ($ per tonne)	287	70–552
Pyro-6F methane emissions (kg per tonne of dry matter)	2.9	2.2–6.4
Avoided rice husk decay methane emissions (kg per tonne of dry matter)	0.33	0.33–2.46
Duration of biochar's effect (years)	50	1–100
Biochar price ($ per tonne)	200	100–300
Pyro-6F production time (% of capacity)	50	25–100
Biochar transportation distance (km)	10	1–25
Soil nitrous oxide emissions (%)	0	−50% to +50

Source: World Bank.

Table B.25 Sensitivity Results for Stable Carbon Content of Biochar Parameter

Stable carbon content of biochar (%)	GHG (tonnes CO₂e per tonne of dry matter)	Change (%)
0	+0.03	−108
50	−0.25	−39
80	−0.41	0
90	−0.47	+15

Source: World Bank.
Note: GHG = greenhouse gas; CO_2e = carbon dioxide equivalent.

parameters that were tested are the fraction of recalcitrant carbon in the biochar, the yield response of onion crops with biochar additions, the price the farmer receives for onions, methane emissions from the pyrolysis unit, methane emissions from avoided rice husk decay, the duration of biochar's agronomic effectiveness, the price of the biochar, the production time for the pyrolysis unit, the biochar transportation distance, and soil nitrous oxide emissions. In order to facilitate the data interpretation, the discussion of each sensitivity test is followed by a list of the results for those impact categories (GHG or monetary value) that are relevant to the test (that is, results are not presented if the value does not vary from the baseline).

The stable carbon content is varied in the range 0–90 percent with a baseline of 80 percent (table B.25). A change in the stable fraction of carbon in the biochar from the baseline of 80 percent to 50 percent lowers the net GHG reductions by 39 percent. Meanwhile, a 90 percent stable carbon fraction increases the net GHG by 15 percent compared to the baseline. If the stable fraction of the carbon in the biochar is 0 percent, the net GHG goes slightly positive, to +0.03 tonnes of CO_2e per tonne of dry matter. From this analysis it is evident that the stable carbon content of the biochar is a major driver in the net GHG balance of the Senegal biochar system.

Table B.26 Sensitivity Results for Yield Response Parameter

Yield response (% of baseline)	Surplus onion (kg per tonne of dry matter)	Change (%)	Revenue ($ per tonne of dry matter)	Change (%)
1-crop effect				
−50	−227	−197	−109	−554
0	0	−100	−43	−279
52	235	0	24	0
233	1,060	+351	261	+988
50-year effect				
−50	−22,700	−195	−6,565	−198
0	0	−100	−43	−100
52	24,000	0	6,696	0
233	106,000	+342	30,394	+354

Source: World Bank.

In order to consider the effect of varying the crop yield response with biochar amendments, a 1-crop biochar effect is compared to a 50-year biochar effect in table B.26. The onion yield is ranged from −50 percent (a decrease in yield with biochar additions as compared to the control) to a 233 percent increase in yield (the maximum percent increase for maize trials).[3] At the baseline value of 52 percent, there is a 235-kilogram surplus of onions per tonne of dry feedstock per crop, or a 24,000-kilogram surplus per tonne of dry matter over 50 years. For the maximum yield increase of 233 percent, the net revenue increases to +$261 per tonne of dry matter and a surplus of 1,060 kilograms of onions per tonne of dry matter for a 1-crop effect, while a 50-year effect results in $30,394 and 106,000 kilograms of onions per tonne of dry matter. With no change in crop yield, the net revenue becomes negative for both the 1-crop and 50-year effects, at −$43. A 50 percent decrease in yield results in about a 200 percent decrease in yield for both the 1-crop and 50-year biochar effects, and the net revenue is dominated by the lost onion sales. The changes in GHG emissions with a variation in crop yield are less than a fraction of a percent from the baseline because the effect of yield increase on soil organic carbon is minimal, as discussed in other case studies.

Onion prices are ranged from $70 to $552 per tonne, with a baseline of $287 per tonne. The lower price is based on the minimum farmer price for onions in Kenya (Weinberger and Pichop 2009), and the upper price assumes the farmer receives the entire retail price for onions in Senegal (David-Benz, Wade, and Egg 2005). The sensitivity analysis compares varying the onion price for the 50-year effect for biochar on crop yields. As the numbers in table B.27 demonstrate, even though the net revenue decreases by 76 percent at the lowest onion price, the net revenues are still significant when considering biochar's 50-year effect ($1,600 per tonne of dry matter) because of the high yield of onions. (Note: there are two onion crops per year, thus a 50-year effect is for 100 onion crops.)

Table B.27 Sensitivity Results for Onion Price Parameter Assuming a 50-Year Effect

Onion price ($ per tonne)	Revenue ($ per tonne of dry matter)	Change (%)
70	1,600	−76
287	6,696	0
552	12,917	+93

Source: World Bank.

Table B.28 Sensitivity Results for Pyro-6F Methane Emissions Parameter

Pyro-6F methane emissions (kg methane per tonne of dry matter)	GHG (tonnes CO_2e per tonne of dry matter)	Change (%)
2.2	−0.43	+5
2.9	−0.41	0
6.4	−0.33	−20

Source: World Bank.
Note: GHG = greenhouse gas; CO_2e = carbon dioxide equivalent.

The methane emissions during pyrolysis were varied from 2.2 to 6.4 kilograms of methane per tonne of dry fuel, based on low and high values found in the literature, where the baseline value is 2.9 kilograms per tonne of dry fuel (table B.28). The low value of 2.2 kilograms of methane is based on the lower range for a controlled continuous process (Brown 2009), while the high of 6.4 kilograms of methane is for an open fire (Johnson et al. 2008). The net GHG reductions change by 5 percent and −20 percent for the lower and upper bounds of the emissions, respectively. This indicates that the methane emissions do not dominate the net GHG emissions of the project life cycle within the reasonable range for the technology. However, although unlikely, under very poor operating conditions with methane emissions as high as an open fire, the impacts could be quite significant on the net GHG balance.

The low end of the emissions range of 0.33 kilograms of methane per tonne of dry matter for rice husk decay is based on calculations following the Clean Development Mechanism (CDM) methodology as discussed in the process description in chapter 5 (table B.29). The upper end of 2.46 kilograms of methane per tonne of dry matter is from emissions during composting of yard waste residues (Hellebrand 1998). A higher methane emissions value for avoided decay results in higher avoided methane emissions, thus a 15 percent increase in net GHG reductions. Similar to the Pyro-6F emissions, the avoided rice husk decay methane emissions do not dominate the GHG balance within the uncertainty range.

The duration of biochar's effect on crop productivity is varied from 1 to 100 years, where the baseline is assumed to be 50 years (table B.30). The "surplus onions" are the quantity of onions grown in excess of what would be grown without biochar additions over the 1-, 50-, or 100-year duration on the area to which biochar was applied by the farmer. This assumes that the biochar is

Table B.29 Sensitivity Results for Rice Husk Decay Methane Emissions Parameter

Rice husk decay methane emissions (kg per tonne of dry matter)	GHG (tonnes CO₂e per tonne of dry matter)	Change (%)
0.33	−0.41	0
2.46	−0.47	+15

Source: World Bank.
Note: GHG = greenhouse gas; CO₂e = carbon dioxide equivalent.

Table B.30 Sensitivity Results for Duration of Biochar's Agronomic Effect Parameter

Duration of biochar's effect on productivity (yr)	Surplus onion (kg per tonne of dry matter)	Revenue ($ per tonne of dry matter)	Change (%)	GHG (tonnes CO₂e per tonne of dry matter)
1	470	91	−99	−0.36
50	23,485	6,696	0	−0.41
100	46,970	13,436	+101	−0.47

Source: World Bank.
Note: GHG = greenhouse gas; CO₂e = carbon dioxide equivalent.

Table B.31 Sensitivity Results for Biochar Price Parameter

Biochar price ($ per tonne)	Revenue ($ per tonne of dry matter)	Change (%)
100	6,618	−1
200	6,696	0
300	6,774	+1

Source: World Bank.

applied at the specified rate of 10 tonnes per hectare. The $287 per tonne value is assigned to the onion surplus. It is important to note that these revenues are summed over the duration of biochar's effect on crop productivity (from as short as 1 year to as long as 100 years). Note also that there are two cropping seasons per year. The net revenue changes by about ±100% over the range of tested timeframes for biochar's agronomic effect.

The price of biochar is varied ±50 percent from $100 to 300 per tonne with a baseline of $200 per tonne (table B.31). The change in net revenues is small, at only ±1 percent, because the relative contribution of the biochar price in the net economics is only 1 percent.

The production time that the Pyro-6F is in operation is ranged from 25 percent to 100 percent of capacity, with a baseline of 50 percent (the production time also is representative of the biomass throughput) (table B.32). Neither the net revenue nor the net GHG change significantly when the production time is varied within this range. The pyrolysis emissions are small compared to the net GHG balance and the economics are dominated by onion revenues, which make

Table B.32 Sensitivity Results for Pyro-6F Production Time Parameter

Pyro-6F production time (% of capacity)	Revenue ($ per tonne of dry matter)	Change (%)	GHG (tonnes CO_2e per tonne of dry matter)
25	6,590	−2	−0.411
50	6,696	0	−0.414
100	6,750	+1	−0.415

Source: World Bank.
Note: GHG = greenhouse gas; CO_2e = carbon dioxide equivalent.

Table B.33 Sensitivity Results of Revenue for Biochar Transportation Distance Parameter

Biochar transport distance (km)	Revenue ($ per tonne of dry matter)	Change (%)
1	6,707	+0.2
10	6,696	0
25	6,679	−0.3

Source: World Bank.

Table B.34 Sensitivity Results for Soil Nitrous Oxide Parameter

Soil nitrous oxide emissions (% of baseline)	GHG (tonnes CO_2e per tonne of dry matter) 1-crop effect	Change (%)	GHG (tonnes CO_2e per tonne of dry matter) 50-year effect	Change (%)
−50	−0.366	+2	−1.12	+170
0	−0.359	0	−0.414	0
50	−0.352	−2	+0.291	−170

Source: World Bank.
Note: GHG = greenhouse gas; CO_2e = carbon dioxide equivalent.

savings from increased production time appear small. However, it is important to recall that the net revenues are calculated per tonne of dry feedstock. The net revenues for the nongovernmental organization (NGO) operating the pyrolysis unit would be affected much differently than those per functional unit or per farmer. Although there may be feedback in terms of biochar price as a function of the production time, at the same time the biochar price may be limited by what farmers can afford to pay.

The transportation distance of the biochar is ranged from 1 to 25 kilometers, with a baseline of 10 kilometers (table B.33). Even at a distance of 25 kilometers, the net revenue decreases by only 0.3 percent for the 50-year biochar effect, corresponding to the increased transportation cost. Thus, the cost of the biochar transportation plays only a small role in the net revenues for the Senegal system.

Soil nitrous oxide emissions are ranged from +50 percent to −50 percent, estimated from literature values (Singh et al. 2010), assuming a nitrogen fertilizer application rate of 30 kilograms of nitrogen per hectare (table B.34). This analysis compares the 1-crop and the 50-year biochar effects. The changes in net GHG reductions are ±2 percent when considered on a 1-crop basis, demonstrating that biochar's effect on soil nitrous oxide emissions are minimal in the short term. However, for a 50-year biochar effect basis, the cumulative soil nitrous oxide emissions could play a more substantial role at ±170 percent of the net GHG of the baseline. The change in soil nitrous oxide emissions as a result of biochar's interaction with nitrogen fertilizer applications is summed over the 50 years with two crops per year, and becomes a very important parameter at this scale. These results, along with the other case studies, emphasize the need for improved data on biochar's role in soil nitrous oxide emissions.

Biographies: Biochar Guidance Group Members

Pedro Sanchez is the Director of the Tropical Agriculture and the Rural Environment Program, Senior Research Scholar, and Director of the Millennium Villages Project at the Earth Institute at Columbia University. He also directs AfSIS, the African Soils Information Service developing the digital soils map of the world. Sanchez was Director-General of the World Agroforestry Centre (ICRAF) headquartered in Nairobi, Kenya, from 1991 to 2001, and served as Co-chair of the United Nations Millennium Project Hunger Task Force. He is also Professor Emeritus of Soil Science and Forestry at North Carolina State University. Sanchez received his BS, MS, and PhD degrees in Soil Science from Cornell University. His professional career has been dedicated to helping eliminate world hunger and absolute rural poverty while protecting and enhancing the tropical environment. He is the author of *Properties and Management of Soils of the Tropics* (rated among the top 10 best-selling books in soil science worldwide) and coauthor of *Halving Hunger: It Can Be Done*, and has authored or coauthored over 250 scientific publications. He is a Fellow of the American Academy of Arts and Sciences, the American Society of Agronomy, the Soil Science Society of America, and the American Association for the Advancement of Science. He serves on the Board of Agriculture and Natural Resources of the National Academy of Sciences. Sanchez has received honorary Doctor of Science degrees from the Catholic University of Leuven, Belgium, the University of Guelph, Canada, and Ohio State University, United States. He has received decorations from the governments of Colombia and Peru, and was anointed Luo Elder with the name of Odera Akang'o by the Luo community of western Kenya. Sanchez is the 2002 World Food Prize Laureate and 2004 MacArthur Fellow.

Rosina M. Bierbaum serves as Dean, School of Natural Resources and Environment, and Professor, Natural Resources and Environmental Policy, University of Michigan. She was appointed Dean of the School of Natural Resources and Environment in October 2001. Previously, she served in environmental science policy leadership positions in both the legislative and executive branches of U.S. government, culminating in appointment as Director of the Environment Division of the White House Office of Science and Technology

Policy, a Senate-confirmed position. In April 2008, Dr. Bierbaum was selected by the World Bank to codirect the 2010 edition of the *World Development Report*, an annual publication that focuses on a different topic each year and aims both to consolidate existing knowledge on a particular aspect of development and to stimulate debate on new directions for development policy. Dr. Bierbaum has been elected a Fellow of the American Academy of Arts and Sciences and the American Association for the Advancement of Science, and was appointed by President Barack Obama to the President's Council of Advisors on Science and Technology. She is currently on the boards of the Federation of American Scientists, the Environmental and Energy Study Institute, the Energy Foundation, and the Gordon and Betty Moore Foundation. She is also a member of the Executive Committee for the Tyler Prize for Environmental Achievement. Dr. Bierbaum received her BS in Biology and BA in English from Boston College, and earned her PhD in Ecology and Evolution at the State University of New York, Stony Brook.

Sasha Lyutse is a Policy Analyst at the Center for Market Innovation of the Natural Resources Defense Council (NRDC), focusing on climate and energy policy in the agricultural sector. Prior to joining NRDC, she worked as an analyst at Goldman Sachs, providing client relationship management, custodial, financing, and reporting services for hedge funds, as well as at United States embassies in both London and Paris. She holds a Bachelor of Science in Foreign Service from Georgetown University and a dual Masters in Public Administration from the London School of Economics and Sciences Po University in Paris.

Patricia (Pipa) Elias is a forest science and policy consultant. Over the past few years she has represented the Union of Concerned Scientists at the United Nations climate negotiations, working to create science-based reducing emissions from deforestation and forest degradation (REDD)+ policies. She has a bachelor's degree in Environmental Science from the University of Notre Dame and a master's degree in Forestry from Virginia Tech. In the past she has worked for the U.S. Forest Service and conducted soil chemistry research in the Appalachian forests.

Dr. Christian Witt is Senior Program Officer, Agricultural Development, at The Bill & Melinda Gates Foundation. He manages the foundation's subinitiative on soil health, which aims to increase soil fertility and productivity in smallholder farming. Prior to joining the foundation, he was a scientist and project manager in international agricultural research, with more than 15 years' experience in plant nutrition and soil nutrient management in the tropics. Witt spent most of his career working in Asia. Most recently, he was Director of the International Plant Nutrition Institute Southeast Asia, based in Singapore and Malaysia for six years, after a long career at the International Rice Research Institute in the Philippines, where he started his PhD research in 1993 and last held the position of Affiliate Scientist in 2003. A native of Germany, he holds a Diploma (MSc degree) in Biology from the University of Hamburg (1992) and a Doctorate degree in Biology from Justus-von-Liebig University, Germany, received in 1997. Dr. Witt has coordinated multinational research projects working on rice, maize, wheat, and oil palm, published more than 100 papers, including 25 refereed journal articles and book chapters, served as book editor, and developed a wide range of training materials and extension tools, including pocket guides, software and film products, and a leaf color chart for efficient nitrogen management in rice. He is codeveloper of the widely accepted site-specific nutrient management (SSNM) approaches for rice and maize in Asia.

References

Abhayawick, L., J. C. Laguerre, V. Tauzin, and A. Duquenoy. 2002. "Physical Properties of Three Onion Varieties as Affected by the Moisture Content." *Journal of Food Engineering* 55: 253–62.

Aggarwal, R. K., and S. S. Chandel. 2004. "Review of Improved Cookstoves Programme in Western Himalayan State of India." *Biomass and Bioenergy* 27: 131–44.

Ansems, L., and T. N. Ligthart. 2002. "Data Certification for LCA Comparisons: Inventory of Current Status and Strength and Weakness Analysis." TNO Report R2002-601, TNO, The Netherlands.

Antal, M. J., K. Mochidzuki, and L. S. Paredes. 2003. "Flash Carbonization of Biomass." *Industrial and Engineering Chemistry Research* 42: 3690–99.

Asai, H., B. K. Samson, H. M. Stephan, K. Songyikhangsuthor, K. Homma, Y. Kiyono, Y. Inoue, T. Shiraiwa, and T. Horie. 2009. "Biochar Amendment Techniques for Upland Rice Production in Northern Laos: 1. Soil Physical Properties, Leaf SPAD and Grain Yield." *Field Crops Research* 111 (1–2): 81–84.

Ayodele, A., P. Oguntunde, A. Joseph, and D. M. de Souza. 2009. "Numerical Analysis of the Impact of Charcoal Production on Soil Hydrological Behavior, Runoff Response, and Erosion Susceptibility." *Revista Brasilieria de Ciência do Solo* 33: 137–45.

Azcón, R., and R. M. Tobar. 1998. "Activity of Nitrate Reductase and Glutamine Synthetase in Shoot and Root of Mycorrhizal *Allium cepa*: Effect of Drought Stress." *Plant Science* 133: 1–8.

Baldock, J. A., and R. J. Smernik. 2002. "Chemical Composition and Bioavailability of Thermally Altered *Pinus resinosa* (Red Pine) Wood." *Organic Geochemistry* 33: 1093–109.

Barnes, D., K. Openshaw, K. R. Smith, and R. van der Plas. 1993. "The Design and Diffusion of Improved Cooking Stoves." *World Bank Research Observer* 8 (2): 119–41.

Batjes, N. H. 1996. "Total Carbon and Nitrogen in the Soils of the World." *European Journal of Soil Science* 47: 151–163.

Baumann, H., and A.-M. Tillman. 2004. *The Hitch Hiker's Guide to LCA: An Orientation in Life Cycle Assessment Methodology and Application.* Lund, Sweden: Studentlitteratur AB.

Beagle, E. C. 1978. "Rice Husk Conversion to Energy." FAO Agricultural Services Bulletin 31. Food and Agriculture Organization of the United Nations, Rome.

Beck, D. A., G. R. Johnson, and G. A. Spolek. 2011. "Amending Greenroof Soil with Biochar to Affect Runoff Water Quantity and Quality." *Environmental Pollution* 59 (8–9): 2111–18.

Bird, M. I., C. Moyo, E. M. Veenedaal, J. Lloyd, and P. Frost. 1999. "Stability of Elemental Carbon in a Savanna Soil." *Global Biogeochemical Cycles* 13 (4): 923–32.

Blackwell, P., G. Riethmuller, and M. Collins. 2009. "Biochar Application to Soil." In *Biochar for Environmental Management: Science and Technology*, edited by J. Lehmann and S. Joseph, 207–26. London: Earthscan.

Bond, W. J., and J. E. Keeley. 2005. "Fire as a Global 'Herbivore': The Ecology and Evolution of Flammable Ecosystems." *Trends in Ecology and Evolution* 20 (7): 387–94.

Borlaug, N. 2007. "Feeding a Hungry World." *Science* 318 (5849): 359.

Bradford, S. A., E. Segal, W. Zheng, Q. Q. Wang, and S. S. Hutchins. 2008. "Reuse of Concentrated Animal Feeding Operation Wastewater on Agricultural Lands." *Journal of Environmental Quality* 37 (5): S97–S115.

Bridgwater, A. 2007. "IEA Bioenergy Update 27: Biomass Pyrolysis." *Biomass and Bioenergy* 31: I–V.

Brock, C. A., J. Cozic, R. Bahreini, K. D. Froyd, A. M. Middlebrook, A. McComiskey, J. Brioude, O. R. Cooper, A. Stohl, K. C. Aikin, J. A. D. Gouw, D. W. Fahey, R. A. Ferrare, R.-S. Gao, W. Gore, J. S. Holloway, G. Hübler, A. Jefferson, D. A. Lack, S. Lance, R. H. Moore, D. M. Murphy, A. Nenes, P. C. Novelli, J. B. Nowak, J. A. Ogren, J. Peischl, R. B. Pierce, P. Pilewskie, P. K. Quinn, T. B. Ryerson, K. S. Schmidt, J. P. Schwarz, H. Sodemann, J. R. Spackman, H. Stark, D. S. Thomson, T. Thornberry, P. Veres, L. A. Watts, C. Warneke, and A. G. Wollny. 2011. "Characteristics, Sources and Transport of Aerosols Measured in Spring 2008 during the Aerosol, Radiation and Cloud Processes Affecting Arctic Climate (ARCPAC) Project." *Atmospheric Chemistry and Physics* 11: 2423–53.

Brodowski, S. B. 2004. "Origin, Function, and Reactivity of Black Carbon in the Arable Soil Environment." Unpublished doctoral thesis, University of Bayreuth, Bayreuth, Germany.

Brown, R. C. 2009. "Biochar Production Technology." In *Biochar for Environmental Management: Science and Technology*, edited by J. Lehmann and S. Joseph, 127–46. London: Earthscan.

Brown, R. A., A. Kercherb, T. Nguyen, D. Nagle, and W. Ball. 2006. "Production and Characterization of Synthetic Wood Chars for Use as Surrogates for Natural Sorbents." *Organic Geochemistry* 37 (3): 321–33.

Bruun, S., T. El-Zahery, and L. Jensen. 2009. "Carbon Sequestration with Biochar: Stability and Effect on Decomposition of Soil Organic Matter." *IOP Conference Series: Earth and Environmental Science* 6: 24.

Busscher, W. J., J. M. Novak, D. E. Evans, D. W. Watts, M. A. S. Niandou, and M. Ahmedna. 2010. "Influence of Pecan Biochar on Physical Properties of a Norfolk Loamy Sand." *Soil Science* 175 (1): 10–14.

Carbon Gold. 2009. "General Methodology for Quantifying the Greenhouse Gas Emission Reductions from the Production and Incorporation of Soil of Biochar in Agricultural and Forest Management Systems." http://v-c-s.org/sites/v-c-s.org/files/Methodology%20for%20Biochar%20Production%20and%20Incorporation%20in%20ALM%20and%20IFM.pdf.

Chan, K. Y., L. van Zwieten, I. Meszaros, A. Downie, and S. Joseph. 2007. "Agronomic Values of Greenwaste Biochar as a Soil Amendment." *Soil Research* 45: 629–34.

Chan, K. Y., and Z. Xu. 2009. "Biochar: Nutrient Properties and Their Enhancement." In *Biochar for Environmental Management: Science and Technology*, edited by J. Lehmann and S. Joseph, 67–84. London: Earthscan.

Cheng, C.-H., J. Lehmann, and M. H. Engelhard. 2008. "Natural Oxidation of Black Carbon in Soils: Changes in Molecular Form and Surface Charge along a Climosequence." *Geochimica et Cosmochimica Acta* 72 (6): 1598–610.

Cheng, C.-H., J. Lehmann, J. E. Thies, and S. D. Burton. 2008. "Stability of Black Carbon in Soils across a Climatic Gradient." *Journal of Geophysical Research* 113: G02027.

Cheng, C.-H., J. Lehmann, J. E. Thies, S. D. Burton, and M. H. Engelhard. 2006. "Oxidation of Black Carbon by Biotic and Abiotic Processes." *Organic Geochemistry* 37: 1477–88.

ClimateCare. 2010. "Indicative Programme, Baseline, and Monitoring Methodology for Improved Cook-Stoves and Kitchen Regimes." CDM Gold Standard. http://www.cdmgoldstandard.org/wp-content/uploads/2011/11/GS_Methodology_Cookstove.pdf.

Clough, T. J., F. M. Kelliher, R. R. Sherlock, and C. D. Ford. 2004. "Lime and Soil Moisture Effects on Nitrous Oxide Emissions from a Urine Patch." *Soil Science Society of America Journal* 68: 1600–1609.

Couto, L. 2010. "Brazilian Network of Biomass for Energy (RENABIO)." Presentation at Third International Biochar Conference, Rio de Janeiro, Brazil, September 12–15, 2010.

Crane-Droesch, A., S. Abiven, M. Schmidt, and M. Torn. 2010. "A Meta-Analysis of Plant Response to Biochar." Presentation at Third International Biochar Conference, Rio de Janeiro, Brazil, September 12–15, 2010.

CSIRO (Commonwealth Scientific and Industrial Research Organisation). 2010. "Investigating Biochar: From Source to Sink." http://www.csiro.au/science/Biochar-Overview.html.

David-Benz, H., I. Wade, and J. Egg. 2005. "Market Information and Price Instability: An Insight into Vegetable Markets in Senegal." *EconWPA* 05122005.

De Capitani, E. M., E. Algranti, M. Z. Aantonieta, A. M. Z. Handar, M. A. Albina, A. M. A. Altemani, R. G. Ferreira, A. B. Balthazar, E. M. F. P. Cerqueira, and J. S. Ota. 2007. "Wood Charcoal and Activated Carbon Dust Pneumoconiosis in Three Workers." *American Journal of Industrial Medicine* 50: 191–96.

Defries, R. S., T. Rudel, M. Uriarte, and M. Hansen. 2010. "Letter: Deforestation Driven by Urban Population Growth and Agricultural Trade in the Twenty-First Century." *Nature Geosciences* 3: 178–81.

De Gryze, S., M. Cullen, and L. Durschinger. 2010. "Evaluation of the Opportunities for Generating Carbon Offsets from Soil Sequestration of Biochar." Terra Global Capital, Paper commissioned by the Climate Action Reserve.

DeLuca, T. H., M. D. MacKenzie, M. J. Gundale, and W. E. Holben. 2006. "Wildfire-Produced Charcoal Directly Influences Nitrogen Cycling in Ponderosa Pine Forests." *Soil Science Society of America Journal* 70: 448–53.

Duguma, L. A., I. Darnhofer, and H. Hager. 2009. "The Financial Return of Cereal Farming for Smallholder Farmers in the Central Highlands of Ethiopia." *Experimental Agriculture* 46 (2): 137–53.

Dünisch, O., V. C. Lima, G. Seehann, J. Donath, V. R. Montola, and T. Schwarz. 2007. "Retention Properties of Wood Residues and Their Potential for Soil Amelioration." *Wood Science and Technology* 41 (2): 169–89.

Elad, Y., D. Rav David, Y. Meller Harel, M. Borenshtein, H. Ben Kalifa, A. Silber, and E. R. Graber. 2010. "Induction of Systemic Resistance in Plants by Biochar, a Soil-Applied Carbon Sequestering Agent." *Phytopathology* 100: 913–21.

Elmer, W. H., and J. J. Pignatello. 2011. "Effect of Biochar Amendments on Mycorrhizal Associations and *Fusarium* Crown and Root Rot of Asparagus in Replant Soils." *Plant Disease* 95 (8): 960–66.

Energy and Mining Sector Board. 2001. *The World Bank Group's Energy Program: Poverty Reduction, Sustainability, and Selectivity.* Washington, DC: World Bank.

Erb, K.-H., F. Krausmann, V. Gaube, S. Gingrich, A. Bondeau, M. Fischer-Kowalski, and H. Haberl. 2009. "Analyzing the Global Human Appropriation of Net Primary Production: Processes, Trajectories, Implications—An Introduction." *Ecological Economics* 69 (2): 250–59.

Fairhead, J., and M. Leach. 2009. "Amazonian Dark Earths in Africa?" Chapter 13 of *Amazonian Dark Earths: Wim Sombroek's Vision*, edited by W. I. Woods, W. G. Teixeira, J. Lehmann, C. Steiner, A. M. G. A. WinklerPrins, and L. Rebellato. Springer Science.

FAO (Food and Agriculture Organization of the United Nations). 2008. *Gender and Equity Issues in Liquid Biofuels Production: Minimizing the Risks to Maximize the Opportunities.* Rome: FAO.

———. 2009. *How to Feed the World in 2050.* Rome: FAO.

———. 2010. *Making Integrated Food-Energy Systems Work for People and Climate.* Rome: FAO.

Fisher, B. 2010. "Letter to the Editor: African Exception to Drivers of Deforestation." *Nature Geosciences* 3: 375–76.

Forbes, M. S., R. J. Raison, and J. O. Skjemstad. 2006. "Formation, Transformation and Transport of Black Carbon (Charcoal) in Terrestrial and Aquatic Ecosystems." *Science of the Total Environment* 370: 190–206.

Garcia-Perez, M. 2008. "The Formation of Polyaromatic Hydrocarbons and Dioxins During Pyrolysis: A Review of the Literature with Description of Biomass Composition, Fast Pyrolysis Technologies and Thermochemical Reactions." Washington State University. http://www.pacificbiomass.org/documents/TheFormationOfPolyaromaticHydrocarbonsAndDioxinsDuringPyrolysis.pdf.

Gaunt, J., and A. Cowie. 2009. "Biochar, Greenhouse Gas Accounting and Emissions Trading." In *Biochar for Environmental Management: Science and Technology*, edited by J. Lehmann and S. Joseph, 317–40. London: Earthscan.

Gaunt, J. L., and J. Lehmann. 2008. "Energy Balance and Emissions Associated with Biochar Sequestration and Pyrolysis." *Environmental Science and Technology* 42: 4152–58.

GEF (Global Environment Facility). 2003. *Operational Program on Sustainable Land Management.* Washington, DC: GEF.

———. 2007. "The Global Environment Facility—GEF: Sustainable Forest Management Program." http://www.cifor.cgiar.org/publications/pdf_files/cop/session%204/7-Sumba-4-7-4-GEF's%20Programmatic-NRMT.pdf.

———. 2009. *GEF Sustainable Forest Management & REDD+ Investment Program.* Washington, DC: GEF.

———. 2010. "The GEF Small Grants Program: Community Action, Global Impact." http://sgp.undp.org/.

Glaser, B., E. Balashov, L. Haumaier, G. Guggenberger, and W. Zech. 2000. "Black Carbon in Density Fractions of Anthropogenic Soils of the Brazilian Amazon Region." *Organic Geochemistry* 31 (7–8): 669–78.

Glaser, B., L. Haumaier, G. Guggenberger, and W Zech. 2001. "The 'Terra Preta' Phenomenon: A Model for Sustainable Agriculture in the Humid Tropics." *Naturwissenschaften* 88: 37–41.

Glaser, B., J. Lehmann, and W. Zech. 2002. "Ameliorating Physical and Chemical Properties of Highly Weathered Soils in the Tropics with Charcoal: A Review." *Biology and Fertility of Soils* 35: 219–30.

Golding, C. J., R. J. Smernik, and G. F. Birch. 2004. "Characterisation of Sedimentary Organic Matter from Three South-Eastern Australian Estuaries Using Solid-State[13] C-NMR Techniques." *Marine and Freshwater Research* 55: 285–93.

Graber, E. R., Y. M. Harel, M. Kolton, E. Cytryn, A. Silber, D. R. David, L. Tsechansky, M. Borenshtein, and Y. Elad. 2010. "Biochar Impact on Development and Productivity of Pepper and Tomato Grown in Fertigated Soilless Media." *Plant and Soil* 337: 481–96.

Guggenberger, G., A. Rodionov, O. Shibistova, M. Grabe, O. A. Kasansky, H. Fuchs, N. Mikheyeva, G. Zhazhevskaya, and H. Flessa. 2008. "Storage and Mobility of Black Carbon in Permafrost Soils in the Forest Tundra Ecotone in Northern Siberia." *Global Change Biology* 14: 1367–81.

Haefele, S., C. Knoblauch, M. Gummert, Y. Konboon, and S. Koyama. 2009. "Black Carbon (Biochar) in Rice-Based Systems: Characteristics and Opportunities." In Chapter 26 of *Amazonian Dark Earths: Wim Sombroek's Vision*, edited by W. I. Woods, W. G. Teixeira, J. Lehmann, C. Steiner, A. M. G. A. WinklerPrins, and L. Rebellato. Berlin: Springer Science.

Haefele, S. M., Y. Konboon, W. Wongboon, S. Amarante, A. A. Maarifat, E. M. Pfeiffer, and C. Knoblauch. 2011. "Effects and Fate of Biochar from Rice Residues in Rice-Based Systems." *Field Crops Research* 121 (3): 430–40.

Hamer, U., B. Marschner, S. Brodowski, and W. Amelung. 2004. "Interactive Priming of Black Carbon and Glucose Mineralisation." *Organic Geochemistry* 35 (7): 823–30.

Hammond, J., S. Shackley, S. P. Sohi, and P. Brownsort. 2011. "Prospective Life Cycle Carbon Abatement for Pyrolysis Biochar Systems in the UK." *Energy Policy* 39: 2646–55.

Hayes, M. 2010. "Development of a Biochar Classification System Based on Its Effect on Plant Growth." Presentation at Third International Biochar Conference, Rio de Janeiro, Brazil, September 12–15, 2010.

Hellebrand, H. J. 1998. "Emission of Nitrous Oxide and Other Trace Gases During Composting of Grass and Green Waste." *Journal of Agricultural Engineering Research* 69: 365–75.

Huang, M.-T., and D. L. Ketring. 1987. "Root Growth Characteristics of Peanut Genotypes." *Journal of Agricultural Research of China* 36: 41–52.

Hyman, E. 1985. *The Experience with Improved Charcoal and Wood Stoves for Households and Institutions in Kenya*. Washington, DC: Appropriate Technology International.

IBI (International Biochar Initiative). 2012. "Developing Guidelines for Specifications of Biochars." http://www.biochar-international.org/characterizationstandard.

Iliffe, R. 2009. "Is the Biochar Produced by an Anila Stove Likely to Be a Beneficial Soil Additive?" UKBRC Working Paper 4, United Kingdom Biochar Research Centre.

Ioannidou, O., A. Zabaniotou, E. V. Antonakou, K. M. Papazisi, A. A. Lappas, and C. Athanassiou. 2009. "Investigating the Potential for Energy, Fuel, Materials and Chemicals Production from Corn Residues (Cobs and Stalks) by Non-catalytic and

Catalytic Pyrolysis in Two Reactor Configurations." *Renewable and Sustainable Energy Reviews* 4: 750–62.

IPCC (Intergovernmental Panel on Climate Change). 2006. *2006 IPCC Guidelines for National Greenhouse Gas Inventories.* Japan: IPCC National Greenhouse Gas Inventories Programme, Technical Support Unit.

———. 2007. "Climate Change 2007: Mitigation." Contribution of Working Group III to the *Fourth Assessment Report of the Intergovernmental Panel on Climate Change*, edited by B. Metz, O. R. Davidson, P. R. Bosch, R. Dave, and L. A. Meyer. Cambridge: Cambridge University Press.

ISO (International Organization for Standardization). 1997. "ISO 14040:1997: Environmental Management—Life Cycle Assessment—Principles and Framework." http://www.iso.org/iso/catalogue_detail.htm?csnumber=23151.

Johnson, M., R. Edwards, C. Alatorre Frenk, and O. Masera. 2008. "In-field Greenhouse Gas Emissions from Cookstoves in Rural Mexican Households." *Atmospheric Environment* 42: 1206–22.

Johnson, M., R. Edwards, A. Ghilardi, V. Berrueta, D. Gillen, C. A. Frenk, and O. Masera. 2009. "Quantification of Carbon Savings from Improved Biomass Cookstove Projects." *Environmental Science and Technology* 43 (7): 2456–62.

Jones, M., E. Lopez Capel, and D. Manning. 2008. "Polycyclic Aromatic Hydrocarbons (PAH) in Biochars and Related Materials." Presentation at International Biochar Initiative Conference, Newcastle, UK.

Joseph, S. 2009. "Socio-Economic Assessment and Implementation of Small-Scale Biochar Projects." In *Biochar for Environmental Management: Science and Technology*, edited by J. Lehmann and S. Joseph, 359–74. London: Earthscan.

Kammann, C. I., S. Linsel, J. W. Gößling, and H.-W. Koyro. 2011. "Influence of Biochar on Drought Tolerance of *Chenopodium quinoa* Willd and on Soil-Plant Relations." *Plant and Soil* 345: 195–210.

Kanagawa, M., and T. Nakata. 2007. "Analysis of the Energy Access Improvement and Its Socio-Economic Impacts in Rural Areas of Developing Countries." *Ecological Economics* 62: 319–29.

Kapkiyai, J. J., N. K. Karanja, J. N. Qureshi, P. C. Smithson, and P. L. Woomer. 1999. "Soil Organic Matter and Nutrient Dynamics in a Kenyan Nitisol under Long-Term Fertilizer and Organic Input Management." *Soil Biology and Biochemistry* 31 (13): 1773–82.

Karhu, K., T. Mattila, I. Bergström, and K. Regina. 2011. "Biochar Addition to Agricultural Soil Increased CH_4 Uptake and Water Holding Capacity: Results from a Short-Term Pilot Field Study." *Agriculture, Ecosystems and Environment* 140: 309–13.

Karve, P., R. Prabunhe, S. Shackley, S. Carter, P. Anderson, S. Sohi, A. Cross, S. Haszeldine, S. Haefele, T. Knowles, J. Field, and P. Tanger. 2011. "Biochar for Carbon Reduction, Sustainable Agriculture and Soil Management (BIOCHARM)." Asia Pacific Network for Global Climate Change Research. http://www.apn-gcr.org/newAPN/activities/ARCP/2009/ARCP2009-12NSY-Kerve/ARCP2009-12NSY-Karve.pdf.

Kato, M., D. M. DeMarini, A. B. Carvalho, M. A. V. Rego, A. V. Andrade, A. S. V. Bomfim, and D. Loomis. 2005. "World at Work: Charcoal-Producing Industries in Northeastern Brazil." *Occupational and Environmental Medicine* 62: 128–32.

Kaygusuz, K. 2011. "Energy Services and Energy Poverty for Sustainable Rural Development." *Renewable and Sustainable Energy Reviews* 15: 936–47.

Keiluweit, M., P. S. Nico, M. G. Johnson, and M. Kleber. 2010. "Dynamic Molecular Structure of Plant Biomass-Derived Black Carbon (Biochar)." *Environmental Science and Technology* 44: 1247–53.

Khan, A. A., and D. N. Iortsuun. 1989. "The Pattern of Dry-Matter Distribution during Development in Onion." *Journal of Agronomy and Crop Science* 162: 127–34.

Kimetu, J. M., and J. Lehmann. 2010. "Stability and Stabilisation of Biochar and Green Manure in Soil with Different Organic Carbon Contents." *Australian Journal of Soil Research* 48 (6–7): 577–85.

Kimetu, J. M., J. Lehmann, S. O. Ngoze, D. N. Mugendi, J. M. Kinyangi, S. Riha, L. Verchot, J. W. Recha, and A. N. Pell. 2008. "Reversibility of Soil Productivity Decline with Organic Matter of Differing Quality along a Degradation Gradient." *Ecosystems* 11: 726–39.

Kirimi, L. 2009. "Kenya's Maize Pricing Situation: Challenges and Opportunities." Presentation at Round Table on Kenya's Food Situation: Challenges and Opportunities, Nairobi, September 18, 2009. http://www.tegemeo.org/Maize-pricing-2.asp.

Knicker, H. 2007. "How Does Fire Affect the Nature and Stability of Soil Organic Nitrogen and Carbon? A Review." *Biogeochemistry* 85: 91–118.

Kuzyakov, Y., J. K. Friedel, and K. Stahr. 2000. "Review of Mechanisms and Quantification of Priming Effects." *Soil Biology and Biochemistry* 32 (11–12): 1485–98.

Kuzyakov, Y., I. Subbotina, H. Chen, I. Bogomolova, and X. Xu. 2009. "Black Carbon Decomposition and Incorporation into Soil Microbial Biomass Estimated by 14C Labeling." *Soil Biology and Biochemistry* 41 (2): 210–19.

Laird, D., P. Fleming, B. Wang, R. Horton, and D. Karlen. 2010. "Biochar Impact on Nutrient Leaching from a Midwestern Agricultural Soil." *Geoderma* 158 (3–4): 436–42.

Lal, R. 2009. "Soil Degradation as a Reason for Inadequate Human Nutrition." *Food Security* 1: 45–57.

Lal, R., and D. Pimentel. 2007. "Biofuels from Crop Residues (Editorial)." *Soil and Tillage Research* 93: 237–38.

Lang, T., A. D. Jensen, and P. A. Jensen. 2005. "Retention of Organic Elements during Solid Fuel Pyrolysis with Emphasis on the Peculiar Behavior of Nitrogen." *Energy and Fuels* 19: 1631–43.

Lapola, D. M., R. Schaldach, J. Alcamo, A. Bondeau, J. Koch, C. Koelking, and J. A. Priess. 2010. "Indirect Land-Use Changes Can Overcome Carbon Savings from Biofuels in Brazil." *Proceedings of the National Academy of Sciences of the United States of America* 107 (8): 3388–93.

Lee, J. W., B. Hawkins, D. M. Day, and D. C. Reicosky. 2010. "Sustainability: The Capacity of Smokeless Biomass Pyrolysis for Energy Production, Global Carbon Capture and Sequestration." *Energy and Environmental Science* 3: 1695–1705.

Lehmann, J. 2007. "Bio-Energy in the Black." *Frontiers in Ecology and Environment* 5 (7): 381–87.

———. 2009. "Terra Preta Nova—Where to from Here?" In *Amazonian Dark Earths: Wim Sombroek's Vision*, edited by W. I. Woods, W. G. Teixeira, J. Lehmann, C. Steiner, A. M. G. A. WinklerPrins, and L. Rebellato, 473–486. The Netherlands: Springer Science.

Lehmann, J., C. Czimczik, D. Laird, and S. Sohi. 2009. "Stability of Biochar in Soil." In *Biochar for Environmental Management: Science and Technology*, edited by J. Lehmann and S. Joseph, 183–206. London: Earthscan.

Lehmann, J., J. P. da Silva, M. Rondon, C. M. da Silva, J. Greenwood, T. Nehls, C. Steiner, and B. Glaser. 2002. "Slash-and-Char: A Feasible Alternative for Soil Fertility Management in the Central Amazon?" Paper No. 49, 17th World Congress of Soil Science, Thailand.

Lehmann, J., J. P. da Silva, C. Steiner, T. Nehls, W. Zech, and B. Glaser. 2003. "Nutrient Availability and Leaching in an Archaeological Anthrosol and a Ferralsol of the Central Amazon Basin: Fertilizer, Manure and Charcoal Amendments." *Plant and Soil* 249: 343–57.

Lehmann, J., and S. Joseph. 2009. "Biochar Systems." In *Biochar for Environmental Management: Science and Technology*, edited by J. Lehmann and S. Joseph, 146–68. London: Earthscan.

Lehmann, J., M. Rillig, J. Thies, C.A. Masiello, W.C. Hockaday, and D. Crowley. 2011. "Biochar Effects on Soil Biota: A Review." *Soil Biology and Biochemistry* 43 (9): 1812–36.

Lehmann, J., and M. Rondon. 2006. "Bio-Char Soil Management on Highly Weathered Soils in the Humid Tropics." In *Biological Approaches to Sustainable Soil Systems*, edited by N. Uphoff, 517–30. Boca Raton, FL: CRC Press.

Lehmann, J., J. Skjemstad, S. Sohi, J. Carter, M. Barson, P. Falloon, K. Coleman, P. Woodbury, and E. Krull. 2008. "Australian Climate Carbon Cycle Feedback Reduced by Soil Black Carbon." *Nature Geoscience* 1 (12): 832–35.

Lehmann, J., and D. Solomon. 2010. "Organic Carbon Chemistry in Soils Observed by Synchrotron-Based Spectroscopy." In *Synchroton-Based Techniques in Soils and Sediment*, edited by B. Singh and M. Gräfe, 289–312. Amsterdam: Elsevier.

Liang, B., J. Lehmann, S. P. Sohi, J. E. Thies, B. O'Neill, L. Trujillo, J. Gaunt, D. Solomon, J. Grossman, E. G. Neves, and F. J. Luizão. 2010. "Black Carbon Affects the Cycling of Non-black Carbon in Soil." *Organic Geochemistry* 41 (2): 206–13.

Liang, B., J. Lehmann, D. Solomon, J. Kinyangi, J. Grossman, B. O'Neill, J. O. Skjemstad, J. Thies, F. J. Luizão, J. Petersen, and E. G. Neves. 2006. "Black Carbon Increases Cation Exchange Capacity in Soils." *Soil Science Society of America Journal* 70 (5): 1719–30.

Liang, B., J. Lehmann, D. Solomon, S. Sohi, J. E. Thies, J. O. Skjemstad, F. J. Luizão, M. H. Engelhard, E. G. Neves, and S. Wirick. 2008. "Stability of Biomass-Derived Black Carbon in Soils." *Geochimica et Cosmochimica Acta* 72 (24): 6069–78.

Libra, J. A., K. S. Ro, C. Kammann, A. Funke, N. D. Berge, Y. Neubauer, M.-M. Titirici, C. Fühner, O. Bens, J. Kern, and K. H. Emmerich. 2011. "Hydrothermal Carbonization of Biomass Residuals: A Comparative Review of the Chemistry, Processes and Applications of Wet and Dry Pyrolysis." *Biofuels* 2 (1): 89–124.

Lim, H. H., Z. Domala, S. Joginder, S. J. Lee, C. S. Lim, and C. M. A. Bakar. 1984. "Rice Millers' Syndrome: A Preliminary Report." *British Journal of Industrial Medicine* 41: 445–49.

Lundie, S., and G. M. Peters. 2005. "Life Cycle Assessment of Food Waste Management Options." *Journal of Cleaner Production* 13 (3): 275–86.

MacCarty, N., D. Ogle, D. Still, T. Bond, and C. Roden. 2008. "A Laboratory Comparison of the Global Warming Impact of Five Major Types of Biomass Cooking Stoves." *Energy for Sustainable Development* 12 (2): 56–65.

MacCarty, N., D. Ogle, D. Still, T. Bond, C. Roden, and B. Willson. 2007. *Laboratory Comparison of the Global-Warming Potential of Six Categories of Biomass Cooking Stoves*. Creswell, OR: Aprovecho Research Center.

Major, J., J. Lehmann, M. Rondon, and C. Goodale. 2010. "Fate of Soil-Applied Black Carbon: Downward Migration, Leaching and Soil Respiration." *Global Change Biology* 16: 1366–79.

Manning, D. A. C., and E. Lopez-Capel. 2009. "Test Procedures for Determining the Quantity of Biochar within Soils." In Chapter 17 of *Biochar for Environmental Management: Science and Technology*, edited by J. Lehmann and S. Joseph. London: Earthscan.

Martin, D. A., and J. A. Moody. 2001. "Comparison of Soil Infiltration Rates in Burned and Unburned Mountainous Watersheds." *Hydrological Processes* 15: 2893–903.

Masiello, C. A. 2004. "New Directions in Black Carbon Organic Geochemistry." *Marine Chemistry* 92: 201–13.

Mitra, A., T. S. Bianchi, B. A. McKee, and M. Sutula. 2002. "Black Carbon from the Mississippi River: Quantities, Sources, and Potential Implications for the Global Carbon Cycle." *Environmental Science and Technology* 36: 2296–302.

Moebius-Clune, B. N., H. M. van Es, O. J. Idowu, R. R. Schindelbeck, D. J. Moebius-Clune, and D. W. Wolfe. 2008. "Long-Term Effects of Harvesting Maize Stover and Tillage on Soil Quality." *Soil Science Society of America Journal* 72 (4): 960–69.

Molina, M., D. Zaelke, K. M. Sarma, S. O. Andersen, V. Ramanathan, and D. Kaniaru. 2009. "Reducing Abrupt Climate Change Risk Using the Montreal Protocol and Other Regulatory Actions to Complement Cuts in CO_2 Emissions." *Proceedings of the National Academy of Sciences* 106 (49): 20616–21.

Müller, D., and J. Mburu. 2009. "Forecasting Hotspots of Forest Clearing in Kakamega Forest, Western Kenya." *Forest Ecology and Management* 257: 968–77.

Mungai, N. W., and P. P. Motavalli. 2006. "Litter Quality Effects on Soil Carbon and Nitrogen Dynamics in Temperate Alley Cropping Systems." *Applied Soil Ecology* 31 (1–2): 32–42.

Murphy, L., and P. Edwards. 2003. *Bridging the Valley of Death: Transitioning from Public to Private Sector Financing*. Golden, CO: National Renewable Energy Laboratory.

National Institute for Soils and Fertilizers. 1996. "Data Set." Tu Liem, Hanoi, Vietnam.

Nguyen, B. T., and J. Lehmann. 2009. "Black Carbon Decomposition under Varying Water Regimes." *Organic Geochemistry* 40: 846–53.

Nguyen, B. T., J. Lehmann, W. C. Hockaday, S. Joseph, and C. A. Masiello. 2010. "Temperature Sensitivity of Black Carbon Decomposition and Oxidation." *Environmental Science and Technology* 44 (9): 3324–31.

Nguyen, B. T., J. Lehmann, J. Kinyangi, R. Smernik, S. J. Riha, and M. J. Engelhard. 2008. "Long-Term Black Carbon Dynamics in Cultivated Soil." *Biogeochemistry* 92: 163–76.

Nhantumbo, I., and A. Salomão. 2010. *Biofuels, Land Access, and Rural Livelihoods in Mozambique*. London: International Institute for Environment and Development.

Norberg-Bohm, V. 2002. "Pushing and Pulling Technology into the Marketplace: The Role of Government in Technology Innovation in the Power Sector." In *The Role of Government in Energy Technology Innovation: Insights for Government Policy in the Energy Sector*, edited by V. Norberg-Bohm, 127–46. Energy Technology Innovation Project, Belfer Center for Science and International Affairs, John F. Kennedy School of Government, Harvard University.

Novak, J. M., W. J. Busscher, D. L. Laird, M. Ahmedna, D. W. Watts, and M. A. S. Niandou. 2009. "Impact of Biochar Amendment on Fertility of a Southeastern Coastal Plain Soil." *Soil Science* 174 (2): 105–12.

Novak, J. M., W. J. Busscher, D. W. Watts, D. A. Laird, M. A. Ahmedna, and M. A. S. Niandou. 2010. "Short-Term CO_2 Mineralization after Additions of Biochar and Switchgrass to a Typic Kandiudult." *Geoderma* 154: 281–88.

Offerman, R., T. Seidenberger, D. Thrän, M. Kaltschmitt, S. Zinoviev, and S. Miertus. 2011. "Assessment of Global Bioenergy Potentials." *Mitigation and Adaptation Strategies for Global Change* 16: 103–15.

Ogawa, M., and Y. Okimori. 2010. "Pioneering Works in Biochar Research, Japan." *Australian Journal of Soil Research* 48: 489–500.

Pacala, S., and R. Socolow. 2004. "Stabilization Wedges: Solving the Climate Problem for the Next 50 Years with Current Technologies." *Science* 305 (5686): 968–72.

Pennise, D. M., K. R. Smith, J. P. Kithinji, M. E. Rezende, T. J. Raad, J. F. Zhang, and C. W. Fan. 2001. "Emissions of Greenhouse Gases and Other Airborne Pollutants from Charcoal Making in Kenya and Brazil." *Journal of Geophysical Research—Atmospheres* 106: 24143–55.

Perez, C., C. Roncoli, C. Neely, and J. S. Steiner. 2007. "Can Carbon Sequestration Markets Benefit Low-Income Producers in Semi-arid Africa? Potentials and Challenges." *Agricultural Systems* 94: 2–12.

Phyllis. 2008. "The Composition of Biomass and Waste." Energy Research Centre of the Netherlands. http://www.ecn.nl/phyllis.

Pimentel, D., A. Marklein, M. A. Toth, M. N. Karpoff, G. S. Paul, R. McCormack, J. Kyriazis, and T. Krueger. 2009. "Food Versus Biofuels: Environmental and Economic Costs." *Human Ecology* 37: 1–12.

Pratt, K., and D. Moran. 2010. "Evaluating the Cost-Effectiveness of Global Biochar Mitigation Potential." *Biomass and Bioenergy* 34: 1149–58.

Qadir, S. A., and T. C. Kandpal. 1995. "A Note on the Financial Evaluation of Improved Biomass Cookstoves." *Renewable Energy* 6: 455–57.

Raju, S. P. 1954. *Smokeless Kitchens for the Millions.* Park Town, Madras: Christian Literature Society.

Raveendran, K., A. Ganesh, and K. C. Khilar. 1995. "Influence of Mineral Matter on Biomass Pyrolysis Characteristics." *Fuel* 74 (12): 1812–22.

Reinaud, G. 2010. "Pro-Natura International: Update on Pro-Natura's Work around the World." Presentation at Third International Biochar Conference, Rio de Janeiro, Brazil, September 12–15, 2010.

Ringler, C. 2010. "Climate Change and Hunger: Africa's Smallholder Farmers Struggle to Adapt." *EuroChoices* 9 (3): 16–21.

Roberts, K. G., B. A. Gloy, S. Joseph, N. R. Scott, and J. Lehmann. 2010. "Life Cycle Assessment of Biochar Systems: Estimating the Energetic, Economic, and Climate Change Potential." *Environmental Science and Technology* 44: 827–33.

Robertson, G. P., and P. M. Groffman. 2007. "Nitrogen Transformations." In *Soil Microbiology, Ecology, and Biochemistry*, edited by E. A. Paul, 341–64. New York, NY: Elsevier.

Roden, C. A., T. C. Bond, S. Conway, A. B. O. Pinel, N. MacCarty, and D. Still. 2009. "Laboratory and Field Investigations of Particulate and Carbon Monoxide Emissions from Traditional and Improved Cookstoves." *Atmospheric Environment* 43: 1170–81.

Rondon, M. A., J. Lehmann, J. Ramirez, and M. Hurtado. 2007. "Biological Nitrogen Fixation by Common Beans (*Phaseolus vulgaris* L.) Increases with Bio-Char Additions." *Biology and Fertility of Soils* 43 (6): 699–708.

Rondon, M. A., D. Molina, M. Hurtado, J. Ramirez, J. Lehmann, J. Major, and E. Amezquita. 2006. "Enhancing the Productivity of Crops and Grasses while Reducing Greenhouse Gas Emissions through Bio-Char Amendments to Unfertile Tropical Soils." Presentation at 18th World Congress of Soil Science, July 9–15, 2006, Philadelphia, PA.

Roth, C. 2011. *Micro Gasification: Cooking with Gas from Biomass*. Germany: GIZ HERA.

Roth, C. H., B. Meyer, H. G. Frede, and R. Derpsch. 1988. "Effect of Mulch Rates and Tillage Systems on Infiltrability and Other Soil Physical Properties of an Oxisol in Paraná, Brazil." *Soil and Tillage Research* 11 (1): 81–91.

RSB (Roundtable on Sustainable Biofuels). 2010. "RSB Principles and Criteria for Sustainable Biofuel Production." http://rsb.epfl.ch/files/content/sites/rsb2/files/Biofuels/Version%202/PCs%20V2/10-11-12%20RSB%20PCs%20Version%202.pdf.

Rumpel, C., V. Chaplot, O. Planchon, J. Bernadou, C. Valentin, and A. Mariotti. 2006. "Preferential Erosion of Black Carbon on Steep Slopes with Slash and Burn Agriculture." *Catena* 54: 30–40.

Searchinger, T. D. 2010. "Biofuels and the Need for Additional Carbon." *Environmental Research Letters* 5 (024007): 1–10.

Seifritz, W. 1993. "Should We Store Carbon in Charcoal?" *International Journal of Hydrogen Energy* 18: 405–7.

Shackley, S., and S. Sohi. 2010. "An Assessment of the Benefits and Issues Associated with the Application of Biochar to Soil." Paper commissioned by the UK Government, UK Biochar Research Centre.

Sheehan, J., V. Camobreco, J. Duffield, M. Graboski, and H. Shapouri. 1998. "Life Cycle Inventory of Biodiesel and Petroleum Diesel for Use in an Urban Bus." Report prepared for U.S. Department of Energy's Office of Fuels Development and U.S. Department of Agriculture's Office of Energy, National Renewable Energy Laboratory, Golden, Colorado.

Shinohara, Y., and N. Kohyama. 2004. "Quantitative Analysis of Tridymite and Cristobalite Crystallized in Rice Husk Ash by Heating." *Industrial Health* 42: 277–85.

Singh, B. P., B. Hatton, B. Singh, A. L. Cowie, and A. Kathuria. 2010. "Influence of Biochars on Nitrous Oxide Emission and Nitrogen Leaching from Two Contrasting Soils." *Journal of Environmental Quality* 39 (4): 1224–35.

Slavich, P., H. M. Tam, T. T. Dung, and B. Keen. 2010. "Industry and Investment, NSW: Rice Husk Biochar Improves Fertility of Sandy Soils in Central Coastal Vietnam." Presentation at Third International Biochar Conference, Rio de Janeiro, Brazil, September 12–15, 2010.

Smith, K. R., D. M. Pennise, P. Khummongkol, V. Chaiwong, K. Ritgeen, J. Zhang, W. Panyathanya, R. A. Rasmussen, M. A. K. Khalil, and S. A. Thorneloe. 1999. "Greenhouse Gases from Small-Scale Combustion Devices in Developing Countries: Charcoal-Making Kilns in Thailand." EPA-600/R-99-109, United States Environmental Protection Agency, National Risk Management Research Laboratory, Washington, DC. http://www.epa.gov/nrmrl/pubs/600r99109/600r99109.pdf.

Sohi, S. P., E. Lopez-Capel, R. Bol, and E. Krull. 2010. "A Review of Biochar and Its Use and Function in Soil." *Advances in Agronomy* 105: 47–82.

Sombroek, W., M. L. Ruivo, P. M. Fearnside, B. Glaser, and J. Lehmann. 2003. "Amazonian Dark Earths as Carbon Stores and Sinks." In *Amazonian Dark Earths: Origin, Properties, Management*, edited by J. Lehmann, D. C. Kern, B. Glaser, and W. I. Woods, 125–140. Dordrecht, The Netherlands: Kluwer Academic Publishers.

Spokas, K. 2010. "Review of the Stability of Biochar in Soils: Predictability of O:C Molar Ratios." *Carbon Management* 1 (2): 289–303.

Spokas, K. A., J. M. Baker, and D. C. Reicosky. 2010. "Ethylene: Potential Key for Biochar Amendment Impacts." *Plant and Soil* 333: 443–52.

Spokas, K. A., W. C. Koskinen, J. M. Baker, and D. C. Reicosky. 2009. "Impacts of Woodchip Biochar Additions on Greenhouse Gas Production and Sorption/Degradation of Two Herbicides in a Minnesota Soil." *Chemosphere* 77 (4): 574–81.

Steinbeiss, S., G. Gleixner, and M. Antonietti. 2009. "Effect of Biochar Amendment on Soil Carbon Balance and Soil Microbial Activity." *Soil Biology and Biochemistry* 41: 1301–10.

Steiner, C. 2010. "Biochar Carbon Sequestration in Tropical Land Use Systems." Presentation at Third International Biochar Conference, Rio de Janeiro, Brazil, September 12–15, 2010.

Steiner, C., K. C. Das, N. Melear, and L. Donald. 2010. "Reducing Nitrogen Loss during Poultry Litter Composting Using Biochar." *Journal of Environmental Quality* 39 (4): 1236–42.

Steiner, C., B. Glaser, W. G. Teixeira, J. Lehmann, W. E. H. Blum, and W. Zech. 2008. "Nitrogen Retention and Plant Uptake on a Highly Weathered Central Amazonian Ferralsol Amended with Compost and Charcoal." *Journal of Plant Nutrition and Soil Science* 171 (6): 893–99.

Steiner, C., W. G. Teixeira, J. Lehmann, T. Nehls, J. L. V. de Macedo, W. E. H. Blum, and W. Zech. 2007. "Long-Term Effects of Manure, Charcoal and Mineral Fertilization on Crop Production and Fertility on a Highly Weathered Central Amazonian Upland Soil." *Plant and Soil* 291 (1–2): 275–90.

Taghizadeh-Toosi, A., T. J. Clough, L. M. Condron, R. R. Sherlock, C. R. Anderson, and R. A. Craigie. 2011. "Biochar Incorporation into Pasture Soil Suppresses In Situ Nitrous Oxide Emissions from Ruminant Urine Patches." *Journal of Environmental Quality* 40 (2): 468–76.

Thomas, A. E. 2008. "Creating a Low-Cost, Low-Particulate Emissions Corn Cob Charcoal Grinder for Use in Peru." Unpublished baccalaureate thesis, Massachusetts Institute of Technology, Cambridge, MA.

Tisdall, J. M., and J. M. Oades. 1982. "Organic Matter and Water-Stable Aggregates in Soils." *Journal of Soil Science* 33: 141–63.

Titirici, M. M., A. Thomas, and M. Antonietti. 2007. "Back in the Black: Hydrothermal Carbonization of Plant Material as an Efficient Process to Treat the CO_2 Problem?" *New Journal of Chemistry* 31: 787–89.

Torres, D. 2011. "Biochar Production with Cook Stoves and Use as a Soil Conditioner in Western Kenya." Unpublished doctoral dissertation, Cornell University, Ithaca, NY.

Torres, D., J. Lehmann, P. Hobbs, S. Joseph, and H. Neufeldt. 2011. "Biomass Availability, Energy Consumption and Biochar Production in Rural Households of Western Kenya." *Biomass and Bioenergy* 35 (8): 3537–46.

Tryon, E. H. 1948. "Effect of Charcoal on Certain Physical, Chemical, and Biological Properties of Forest Soils." *Ecological Monographs* 18: 81–115.

UNEP & WMO. 2011. "Integrated Assessment of Black Carbon and Tropospheric Ozone: Summary for Decision Makers." http://www.unep.org/dewa/Portals/67/pdf/Black_Carbon.pdf.

UNFCCC (United Nations Framework Convention on Climate Change). 2006. "Approved Baseline and Monitoring Methodology AM0041: Mitigation of Methane

Emissions in the Wood Carbonization Activity for Charcoal Production." CDM Executive Board 27. http://cdm.unfccc.int/methodologies/DB/B2SCH5WZLQYHTVSHQ4BIADMCBQ1P9U.

———. 2008. "Methodological Tool: Tool for the Demonstration and Assessment of Additionality." CDM Executive Board 39, Annex 10, United Nations, Bonn, Germany.

———. 2009. "Energy Efficiency Measures in Thermal Applications of Nonrenewable Biomass." CDM Executive Board 51, United Nations, Bonn, Germany.

———. 2010. "Clean Development Mechanism Methodology Booklet." https://cdm.unfccc.int/methodologies.

Unger, P. W., B. A. Stewart, J. F. Parr, and R. P. Singh. 1991. "Crop Residue Management and Tillage Methods for Conserving Soil and Water in Semi-arid Regions." *Soil and Tillage Research* 20 (2–4): 219–40.

U.S. Department of State. 2009. "2008 Human Rights Report: Kenya." http://www.state.gov/g/drl/rls/hrrpt/2008/af/119007.htm.

Van Oost, K., T. A. Quine, G. Govers, S. De Gryze, J. Six, J. W. Harden, G. W. McCarty, G. Heckrath, C. Kosmas, J. V. Giraldez, J. R. Marques da Silva, and R. Merckx. 2007. "The Impact of Agricultural Soil Erosion on the Global Carbon Cycle." *Science* 318 (5850): 626–29.

van Zwieten, L., S. Kimber, A. Downie, S. Morris, S. Petty, J. Rust, and K. Y. Chan. 2010. "A Glasshouse Study on the Interaction of Low Mineral Ash Biochar with Nitrogen in a Sandy Soil." *Australian Journal of Soil Research* 48: 569–76.

van Zwieten, L., S. Kimber, S. Morris, K. Y. Chan, A. Downie, J. Rust, S. Joseph, and A. Cowie. 2010. "Effects of Biochar from Slow Pyrolysis of Papermill Waste on Agronomic Performance and Soil Fertility." *Plant and Soil* 327: 235–46.

van Zwieten, L., S. Kimber, S. Morris, A. Downie, E. Berger, J. Rust, and C. Scheer. 2010. "Influence of Biochars on Flux of N_2O and CO_2 from Ferrosol." *Australian Journal of Soil Research* 48: 555–68.

van Zwieten, L., B. Singh, S. Joseph, S. Kimber, A. Cowie, K. Y. Chan. 2009. "Biochar and Emissions of Non-CO_2 Greenhouse Gases from Soil." In *Biochar for Environmental Management: Science and Technology*, edited by J. Lehmann and S. Joseph, 227–50. London: Earthscan.

Verheijen, F. G. A., S. Jeffery, A. C. Bastos, M. van der Velde, and I. Diafas. 2010. "Biochar Application to Soils: A Critical Scientific Review of Effects on Soil Properties, Processes, and Functions." EUR 24099 EN, Office for the Official Publications of the European Communities, Luxembourg.

von Lützow, M., I. Kögel-Knabner, K. Ekschmitt, E. Matzner, G. Guggenberger, B. Marschner, and H. Flessa. 2006. "Stabilization of Organic Matter in Temperate Soils: Mechanisms and Their Relevance under Different Soil Conditions—A Review." *European Journal of Soil Science* 57: 426–45.

Wang, M. 2007. *The Greenhouse Gases, Regulated Emissions, and Energy Use in Transportation (GREET) Model, 1.8b and 2.7*. UChicago Argonne, LLC.

Wardle, D. A., M.-C. Nilsson, and O. Zackrisson. 2008. "Fire-Derived Charcoal Causes Loss of Forest Humus." *Science* 320 (5876): 629.

Weinberger, K., and G. N. Pichop. 2009. "Marketing of African Indigenous Vegetables along Urban and Peri-urban Supply Chains in Sub-Saharan Africa." In *African Indigenous Vegetables in Urban Agriculture*, edited by C. M. Shackleton, M. W. Pasquini, and A. W. Drescher, 225–44. London: Earthscan.

Whitman, T., C. F. Nicholson, D. Torres, and J. Lehmann. 2011. "Climate Change Impact of Biochar Cook Stoves in Western Kenyan Farm Households: System Dynamics Model Analysis." *Environmental Science and Technology* 45 (8): 3687–94.

Whitman, T., S. M. Scholz, and J. Lehmann. 2010. "Biochar Projects for Mitigating Climate Change: An Investigation of Critical Methodology Issues for Carbon Accounting." *Carbon Management* 1 (1): 89–107.

WHO (World Health Organization). 2000. *Addressing the Links between Indoor Air Pollution, Household Energy and Human Health.* Washington, DC: WHO.

———. 2011. "Indoor Air Pollution and Health." Fact Sheet No. 292. http://www.who.int/mediacentre/factsheets/fs292/en/.

Woods, J., S. Hemstock, and W. Burnyeat. 2006. "Bioenergy Systems at the Community Level in the South Pacific: Impacts and Monitoring." *Mitigation and Adaptation Strategies for Global Change* 11: 469–500.

Woolf, D., J. Amonette, F. Street-Perrott, J. Lehmann, and S. Joseph. 2010. "Sustainable Biochar to Mitigate Global Climate Change." *Nature Communications* 1: 56.

World Bank. 2009. "10 Years of Experience in Carbon Finance: Insights from Working with Carbon Markets for Development and Global Greenhouse Gas Mitigation." World Bank, Washington, DC. http://go.worldbank.org/XIG411DYP0.

———. 2011. *Household Cookstoves, Environment, Health, and Climate Change: A New Look at an Old Problem.* Washington, DC: World Bank.

———. 2013. *On Thin Ice: How Cutting Pollution Can Slow Warming and Save Lives.* Washington, DC: World Bank.

Yanai, Y., K. Toyota, and M. Okazaki. 2007. "Effects of Charcoal Addition on N_2O Emissions from Soil Resulting from Rewetting Air-Dried Soil in Short-Term Laboratory Experiments." *Soil Science and Plant Nutrition* 53: 181–88.

Yonemura, S., and S. Kawashima. 2007. "Concentrations of Carbon Gases and Oxygen and Their Emission Ratios from the Combustion of Rice Hulls in a Wind Tunnel." *Atmospheric Environment* 41 (7): 1407–16.

Zhang, J., K. R. Smith, Y. Ma, S. Ye, F. Jiang, W. Qi, P. Liu, M. A. K. Khalil, R. A. Rasmussen, and S. A. Thorneloe. 2000. "Greenhouse Gases and Other Airborne Pollutants from Household Stoves in China: A Database for Emission Factors." *Atmospheric Environment* 34: 4537–49.

Zimmerman, A. R. 2010. "Abiotic and Microbial Oxidation of Laboratory-Produced Black Carbon (Biochar)." *Environmental Science and Technology* 44 (4): 1295–301.

Zimmerman, A. R., B. Gao, and M.-Y. Ahn. 2011. "Positive and Negative Carbon Mineralization Priming Effects among a Variety of Biochar-Amended Soils." *Soil Biology and Biochemistry* 43 (6): 1169–79.

www.ingramcontent.com/pod-product-compliance
Lightning Source LLC
Chambersburg PA
CBHW080609270326
41928CB00016B/2983